TABLE OF CONTENTS

ACKNOWLEDGEMENTS

The Council for International Organizations of Medical Sciences is greatly indebted to the Science Council of Japan, and particularly its President, Dr. Jiro Kondo, for extending to it the use of the Science Council's excellent facilities and staff in Tokyo for its Conference; and to Mr. Kunio Matsuyama, Mayor of Inuyama City, for the hospitable reception of the conference participants in Inuyama.

Thanks to the efforts of Dr. K. Yagi, Chairman of the Conference Organizing Committee, and President of the Conference, CIOMS was assured of the financial means which enabled it to invite as principal speakers and discussion leaders outstanding authorities on genetic technology and ethics from throughout the world.

Special thanks are due to Dr. E. Inouye of the Science Council of Japan, who, as Co-Chairman with Professor A.M. Capron, of the Programme Committee, ensured the smooth and efficient conduct of the Conference.

The co-sponsorship of the Conference on the parts of the World Health Organization and Unesco was highly appreciated.

CIOMS owes a special debt to the authors of the papers, the chairpersons and rapporteurs of the working groups, and the chairpersons of the plenary sessions. We thank also Dr. J. Gallagher for his assistance in the editing and preparation of this volume, and Mrs. K. Chalaby-Amsler for her efficient work in preparing the Conference and this volume.

INTRODUCTION

Conferences of the Council for International Organizations of Medical Sciences (CIOMS) are aimed at creating international and interdisciplinary forums to enable the scientific and lay communities to exchange views on topics of immediate concern, unhampered by administrative, political or other considerations. They discuss not only the scientific and technical basis of new developments in biology and medicine and other related areas, but also their social, economic, ethical, administrative and legal implications. Participants in these conferences are prominent representatives of different fields of medicine, the natural and the social sciences, philosophy, theology, law and health policymaking. This multidisciplinary approach is felt to be the best means of obtaining a comprehensive picture of issues that do not fall within the exclusive domain of any one profession.

Research at present under way on human genome mapping and sequencing presages a new scientific era, perhaps a revolution, in medicine in the twenty-first century. This field is of much topical concern and the initiation of an international dialogue on rights and responsibilities of individuals and societies is very timely.

In the past several decades, biomedical research has progressed rapidly and its findings have been widely applied in health care. Spectacular advances have been made in molecular biology. The techniques of molecular biology are becoming indispensable in many branches of biomedical research, and in none more than in genetics. Today, findings in genetics are increasingly presented in molecular terms, and the number of genes that have been structurally defined increases at a rapid pace. Efforts have begun in the United States of America, Japan and Europe to map and sequence the entire human genome. It seems likely that by the turn of the century most of the 50-100,000 genes in the human genome will have been isolated and identified. This knowledge will be important not only for diagnosing, treating, and even preventing diseases caused by single-gene defects but also for better understanding, and even treatment, of those major common diseases that result from the interaction of genetic predisposition with various environmental factors.

More than 4,300 single-gene disorders have been identified, and in about 10% of these the protein abnormality has been defined. The rapid development of molecular and cell genetics is making it increasingly possible to localize the genes responsible for disease, even if the defective gene product is not known. More than 1,400 genes have now been mapped to a particular chromosome and 40 of these have been defined in depth. They include the genes responsible for cystic fibrosis, hemoglobinopathies, some hemolytic anemias, phenylketonuria, and hemophilia. In addition, rapid progress may make it possible to isolate the genes responsible for Huntington's disease and some muscular diseases, including Duchenne muscular dystrophy. It is calculated that 1% of all liveborn infants in the developed world have single-gene disorders. At present, it is possible to diagnose some

100 genetic disorders by chorionic villus sampling at eight to ten weeks of pregnancy, or by amniocentesis and cell analysis at 16-18 weeks. An increasing number of fetal malformations can also be diagnosed by ultrasound examination. As time passes, many more will be added to the list of disorders that can be diagnosed prenatally, and the techniques will become more precise.

Many common and debilitating diseases are the result of an interaction between multiple genes and environmental risk factors. Examples are cardiovascular diseases, some cancers, diabetes, major mental disorders including Alzheimer's disease, peptic ulcer and rheumatoid arthritis. Within the next decade biomedical research, particularly molecular genetics, is expected to localize the genes responsible for these diseases and to cast light on the interacting environmental factors. This will lead to improved methods of diagnosis, treatment and prevention.

These scientific developments will raise difficult ethical issues. Single-gene disorders will be increasingly diagnosed prenatally and postnatally. To what extent should prenatal screening be implemented? At present, very few single-gene disorders can be treated. Even if gene therapy becomes a reality within the next decade, it is very likely for many years to be limited to a few diseases. Thus, results of prenatal and postnatal screening will be used primarily for genetic counselling or abortion. The ethical issues are many and can be demonstrated most poignantly by Huntington's disease. This disease becomes apparent only in adulthood, typically after the age of 35 years, by which time the patient is likely to have had offspring. It is characterized by neurological disorders, intellectual and emotional impairment, and premature death. Yet it can be diagnosed at an early stage. Should the consequence of this diagnosis be abortion or not? If the child is born, should the information be withheld or disclosed, and if disclosed, when?

Ethical issues are just as important in the case of multifactorial diseases. The possibilities of improved detection of individual predisposition to a given disease must be weighed against the risk of misuse of this information. For example, an employer who knows of an employee's predisposition to a disease may decide to dismiss him or, even more likely, not to hire him. It might be argued, of course, that such a person should not be exposed to particular work-place factors that could trigger disease.

Another ethical issue arises when advances in diagnosis by means of genetic techniques come faster than advances in treatment. What is the use to the individual of knowing of a predisposition to a disease if there is no treatment? Can genetic screening be recommended? Is it enough to justify genetic screening that there are preventive measures to postpone onset of disease or ameliorative measures to lessen its effects?

Other ethical issues include the extent to which research should be controlled, and who should control it; whether misuse of the knowledge acquired can be precluded, and how; how it will be possible to predict the consequences of genetic intervention and to avoid or mitigate the unacceptable; who should own or otherwise benefit commercially from the products of genetic research; how and from whom is consent to be obtained for genetic interventions, particularly if they may affect future generations; and how will privacy and confidentiality be protected.

The main objective of the Conference was to stimulate an international, interdisciplinary and transcultural dialogue on the ethical issues outlined above.

The Conference brought together some 100 participants from 24 countries and a diversity of professions, religions and cultures. They participated in their personal capacity, and not as representatives of organizations, institutions or countries. This created an unhampered atmosphere ideal for open discussion, free from formal declarations or statements. The Conference reached broad agreement on a number of important issues and at its final session, held in Inuyama City, agreed on a declaration, entitled "The Declaration of Inuyama", reproduced on page 1.

This book is arranged according to the programme of the Conference which was organized as four plenary sessions and a series of discussions of three working groups. The first part consists of the opening speeches and the keynote address, which provides the guidelines for the discussion of the main topics. The second part contains the background papers on human genome mapping, genetic screening and gene therapy (or, as some participants suggested, genetic therapy), covering those subjects from scientific, ethical and policymaking points of view. The final part presents the reports of the three working groups. It covers the topics of human genome mapping, genetic screening and human gene therapy and personal reflections on the Conference from the points of view of a scientist and a physician.

Zbigniew Bankowski, M.D.
Secretary-General, CIOMS

THE DECLARATION OF INUYAMA

Human Genome Mapping, Genetic Screening and Gene Therapy

The Council for International Organizations of Medical Sciences held its XXIVth Round Table Conference, on the subject of *Genetics, Ethics and Human Values: Human Genome Mapping, Genetic Screening and Therapy,* in Tokyo and in Inuyama City, Japan, from 22-27 July 1990. The Conference was held under the auspices of the Science Council of Japan, and cosponsored by the World Health Organization and the United Nations Educational, Scientific and Cultural Organization. It was the fifth in a series entitled *Health Policy, Ethics and Human Values: An International Dialogue,* begun in Athens in 1984. The participants, numbering 102, came from 24 countries, representing all continents.

In addition to biomedical scientists and physicians, the participants represented a wide range of disciplines including sociology, psychology, epidemiology, law, social policy, philosophy and theology, and brought with them experience in hospital and public health medicine, universities and private industry, and the executive and legislative branches of government. Through presentations and discussions in plenary sessions and working groups, they reached broad agreement on a number of central issues. At its final session the Conference agreed on the following Declaration.

I. Discussion of human genetics is dominated today by the efforts now under way on an international basis to map and sequence the human genome. Such attention is warranted by the scale of the undertaking and its expected contribution to knowledge about human biology and disease. At the same time, the nature of the undertaking, concerned as it is with the basic elements of life, and the potential for abuse of the new knowledge which the project will generate, are giving rise to anxiety. The Conference agrees that efforts to map the human genome present no inherent ethical problems but are eminently worthwhile, especially as the knowledge revealed will be universally applicable to benefit human health. In terms of ethics and human values, what must be assured are that the manner in which gene mapping efforts are implemented adheres to ethical standards of research and that the knowledge gained will be used appropriately, particularly in genetic screening and gene therapy.

II. Public concern about the growth of genetic knowledge stems in part from the misconception that while the knowledge reveals an essential aspect of humanness it also diminishes human beings by reducing them to mere base pairs of deoxyribonucleic acid (DNA). This misconception can be corrected by education of the public and open discussion, which should reassure the public that plans for the medical use of genetic findings and techniques will be made openly and responsibly.

1

III. Some types of genetic testing or treatment not yet in prospect could raise novel issues — for example, whether limits should be placed on DNA alterations in human germ cells, because such changes would affect future generations, whose consent cannot be obtained and whose best interests would be difficult to calculate. The Conference concludes, however, that for the most part present genetic research and services do not raise unique or even novel issues, although their connection to private matters such as reproduction and personal health and life prospects, and the rapidity of advances in genetic knowledge and technology, accentuate the need for ethical sensitivity in policy-making.

IV. It is primarily in regard to genetic testing that the human genome project gives rise to concern about ethics and human values. The identification, cloning and sequencing of new genes without first needing to know their protein products greatly expand the possible scope for screening and diagnostic tests. The central objective of genetic screening and diagnosis should always be to safeguard the welfare of the person tested: test results must always be protected against unconsented disclosure, confidentiality must be ensured at all costs, and adequate counselling must be provided. Physicians and others who counsel should endeavour to ensure that all those concerned understand the difference between being the carrier of a defective gene and having the corresponding genetic disease. In autosomal recessive conditions, the health of carriers (heterozygotes) is usually not affected by their having a single copy of the disease gene; in dominant disorders, what is of concern is the manifestation of the disease, not the mere presence of the defective gene, especially when years may elapse between the results of a genetic test and the manifestation of the disease.

V. The genome project will produce knowledge of relevance to human gene therapy, which will very soon be clinically applicable to a few rare but very burdensome recessive disorders. Alterations in somatic cells, which will affect only the DNA of the treated individual, should be evaluated like other innovative therapies. Particular attention by independent ethical review committees is necessary, especially when gene therapy involves children, as it will for many of the disorders in question. Interventions should be limited to conditions that cause significant disability and not employed merely to enhance or suppress cosmetic, behavioural or cognitive characteristics unrelated to any recognized human disease.

VI. The modification of human germ cells for therapeutic or preventive purposes would be technically much more difficult than that of somatic cells and is not at present in prospect. Such therapy might, however, be the only means of treating certain conditions, so continued discussion of both its technical and its ethical aspects is essential. Before germ-line therapy is undertaken, its safety must be very well established, for changes in germ cells would affect the descendants of patients.

VII. Genetic researchers and therapists have a strong responsibility to ensure that the techniques they develop are used ethically. By insisting on truly voluntary programmes designed to benefit directly those involved, they can ensure that no precedents are set for eugenic programmes or other misuse of the techniques by the State or by private parties. One means of ensuring the setting and observance of ethical standards is continuous multidisciplinary and transcultural dialogue.

VIII. The needs of developing countries should receive special attention, to ensure that they obtain their due share of the benefits that ensue from the human genome project. In particular, methods and techniques of testing and therapy that are affordable and easily accessible to the populations of such countries should be developed and disseminated whenever possible.

OPENING OF THE CONFERENCE

F. Vilardell
President, Council for International Organizations
of Medical Sciences

Your Excellency, Dr Hiroshi Nakajima, Director-General of WHO, Professor Kondo, President of the Science Council of Japan, Professor Inouye, Professor Yagi, Dr Wyngaarden, Dr Bankowski, Members of CIOMS, Distinguished Guests.

May I welcome all of you to this XXIVth CIOMS Conference, on "Genetics, Ethics and Human Values: Human Genome Mapping, Genetic Screening and Therapy". This Conference is part of a general programme on Health Policy, Ethics and Human Values initiated by CIOMS in 1985, with successive meetings in Athens, Noordwijk (Netherlands), Bangkok and Geneva.

It is now easy to see how the general topic of genetic control is important and what an enormous impact it may have, not only on the science of medicine but also on general political and sociological issues and on religious feelings. The education of the public on these matters cannot be over-emphasized.

How genome mapping, genetic screening and gene therapy will affect individual freedom vis-à-vis the rights of society is one of the many issues that may arise and be discussed and answered in this Conference.

May I take this opportunity to express to Director-General Nakajima our appreciation of his presence at this opening session and of his constant help to CIOMS endeavours. May I thank also Professor Kondo for his hospitality, Dr Inouye for his help, Dr Bankowski for his painstaking preparatory work, and last, but not least, Professor Kunio Yagi, who made this meeting possible.

The XXIVth CIOMS Conference is open.

Jiro Kondo,
President, Science Council of Japan

Mr. Chairman, Distinguished Guests, Ladies and Gentlemen.

It is a great pleasure to have this opportunity, on behalf of the Science Council of Japan, to speak with each of you who have gathered here from more than 20 countries, at the opening of the Conference of the Council for International Organizations of Medical Sciences.

The Science Council of Japan was established in 1949 as an organ representing qualified Japanese scientists both domestically and internationally, to cover all scientific fields in the Cultural, Social and Natural Sciences Divisions. The aim of the Council is to promote scientific development and, through science, to improve the administrative, the industrial and the national life.

Since our founding, we have been working to contribute to the progress of science, in cooperation with the academic world, by sponsoring many international congresses here in Japan, and by delegating Japanese representatives to international congresses held overseas. We do this because we believe that international scientific exchange is one of our most important duties.

On the occasion of the opening of the conference of the Council for International Organizations of Medical Sciences, it is a particular pleasure to have this chance to meet with so many distinguished scientists from all over the world, and to attend your lectures and presentations.

The Council for International Organizations of Medical Sciences, of which the Science Council of Japan is a member, has been contributing to the formulation of international guidelines concerning bioethics and medical ethics by discussing various topics at its International Conferences every year, and publishing the results of new studies.

As you know, life sciences aim at the understanding of the life of human beings as well as all living things by assimilating knowledge obtained from studies in various biological fields, including molecular biology. The recent advances in the life sciences are indeed remarkable. The human genome sequence project has begun in developed countries to make a genetic map of man, and is expected to have a profound influence on numerous research areas. In Japan, many researchers have undertaken this kind of study and are now taking part in such activity throughout the world. I am delighted that so many scientists who are attending this conference will discuss the topics of the human genome, genetic screening and gene therapy from the international viewpoint.

Finally, I sincerely hope for the great success of this conference. I also hope that each of you from abroad will enjoy your stay in Japan. I believe that this conference will become truly memorable for you through personal contacts with fellow scientists, and that you will learn more about Japanese culture.

E. Inouye
Chairman, Conference Programme Committee

Mr. Chairman, Distinguished Participants, Ladies and Gentlemen.

On behalf of the Programme Committee I extend a hearty welcome to all participants in the 25th CIOMS Conference, on Genetics, Ethics and Human Values.

It was in October 1988 that CIOMS officers, Dr Bankowski, Secretary General, and Professor Yagi, Member of the Executive Committee, submitted to the Science Council of Japan a proposal to convene an international and interdisciplinary conference to deliberate upon the issues raised by the remarkable developments in human molecular genetics, cell genetics and allied biomedical sciences over the recent decade. The Seventh Division of the Council and its CIOMS subcommittee agreed to collaborate with CIOMS, and a Programme Committee co-chaired by Professor Capron was organized to formalize the structure of the Conference in accordance with the framework of CIOMS activity.

The subtitle of the Conference, "Human Genome Mapping, Genetic Screening and Therapy", was selected on the grounds that the Human Genome Mapping, which is a central part of the Human Genome Project, should contribute greatly to the advancement of the biomedical sciences, but that at the same time it will have an unprecedented impact upon future human society, and could have profound positive and negative effects upon ourselves.

It is anticipated that, as the Human Genome Project proceeds, a considerable amount of data will be obtained on the genetic organization of humans at cellular and molecular levels. The Project eventually aims at exploring the entire genetic information carried by our genome, and every step towards this objective will contribute more or less to the advancement of the biomedical sciences and human welfare. Many gene loci still unknown but responsible for the etiology of genetic disorders will be identified, leading to diagnosis, prevention and treatment of genetic diseases.

However, we still do not know how much genetic information we have in our genome. We know about 4,000 gene loci but the total number is estimated at no less than 50,000, which represents only 5% of the total genome.

We are now at the starting point of exploring the fundamental design of mankind, which governs not only our physical features, but also most probably, our behavioural characteristics. We may also ask whether it is possible to guarantee to make public the results of the studies while at the same time protecting the privacy and confidentiality of the subjects involved in it.

We shall encounter more difficult problems as the accomplishments of the project become applicable to medical practice. More and more, we shall be able to detect carriers of abnormal genes — that is, people can be diagnosed as to the presence or absence of a variety of abnormal genes, which though not manifesting symptoms themselves can be transmitted

6

to offspring, or which may be at the stage before the onset of disease. If the diseases cannot be treated at the time, under what conditions are we justified in screening individuals or populations for the presence of such abnormal genes?

If the modification of environmental factors does not suffice to prevent the onset of diseases, medical scientists naturally look for other means of prevention and treatment, including artificial alteration of an abnormal genetic constitution. This is gene therapy and it becomes more easily accessible as we acquire the knowledge of molecular basis of human genetic disorders. If such intervention is performed on a human germ-cell line, what is the consequence to human evolution and can we be optimistic about survival of our offspring in the future?

I have mentioned only a few of the issues. Some have already been deliberated upon on several occasions, but the present Conference is particularly notable since so many distinguished participants, with diverse historical and cultural backgrounds, who are also experts on a wide range of human activities, have come from different parts of the world to exchange views on the points at issue. I think that this is critically important and I hope this Conference will prove to be a vital milestone in our decision-making on the future of human welfare.

H. Nakajima
Director-General, World Health Organization

Mr. Chairman, Distinguished Guests and Colleagues, Ladies and Gentlemen.
On behalf of the World Health Organization I should like to greet all of you. My sincere thanks for the invitation to be here today go to Dr Bankowski, the very able and dynamic Secretary-General of CIOMS, and to the local organizers of this Conference, particularly to you, Dr Yagi, our Chairman. I welcome this opportunity to be with you to discuss current problems in areas related to human genetics. I should like to assure you of my full support for the initiative of CIOMS in organizing a broad international debate within the framework of health policy, ethics and human values. The subjects selected for this debate are of great relevance to the further development of medicine and biology, their societal implications, and their benefits to mankind.

Thanks to achievements in genetic and molecular technology, great progress has been made in recent years in understanding gene structure at the level of DNA. The loci for a number of genetic diseases have been identified, and this includes their mapping in chromosomes. This research has been greatly enhanced by the activities of the Human Genome Project, launched by the United States of America, Japan and a number of European countries, and coordinated by the Human Genome Organization, better known as HUGO, as well as Unesco and other interested agencies. This research has practical objectives, namely, elucidation of the molecular basis of diseases, and development of new modalities for their diagnosis and strategies for novel preventive and therapeutic approaches.

We know that genetic factors are important in human health, and that genes can determine a wide spectrum of conditions, ranging from simply inherited single-gene defects to complex genetic predisposition to common diseases. Over 4300 monogenically determined conditions have already been identified, many representing disease loci. Among these, approximately 400 can be precisely mapped to a specific chromosome region. During the next decade or so, complete maps of every chromosome are expected to become available. The nature of inherited diseases, as well as common diseases with a genetic component, such as coronary heart disease, diabetes, hypertension, and certain oncological and mental disorders, will be understood. This has monumental implications for the control of disease, through both prenatal diagnosis and new methods of presymptomatic treatment, before the clinical manifestation of a defective gene.

Because of such technical advances as gene amplification, DNA diagnosis may soon become routine not only in major centres, but also in local laboratories and in general medical practice, since the techniques are relatively simple, rapid and cheap, and transferable to laboratories in developing countries. For single-gene defects, in cases of thalassaemia, sickle-cell disease or cystic fibrosis, there have already been remarkable achievements. There is promise of similar progress for some common diseases.

Thus, progress in human genome mapping will undoubtedly contribute significantly to the successful identification of genes of major importance for human health. However, such progress will depend on extensive patient-oriented studies. Education is necessary to ensure that the public realizes that these new methods of analysis and treatment are generally similar to existing ones, although more powerful.

It is important to recognize that a number of basic human rights are involved. Ethical questions should be viewed in the light of the general objective of these procedures, which is to help people to live long and healthy lives and reproduce as normally and as responsibly as possible.

Health is a fundamental human right. As we enter the 1990s, human rights, social justice and ethical issues are becoming more and more important. Human rights in health and medicine must be seen within the wider framework of basic human rights, for the rights of the patient are the rights of man. They involve three related issues: freedom of expression and the right to be informed; respect for the integrity of the individual, including freedom to decide; and prohibition of discrimination of any kind. In the 1990s I intend to emphasize the human rights aspects of health in all WHO's programme activities.

I shall be selective and mention only a few of the ethical issues involved in the field of medical genetics. The principles that guide clinical geneticists, which are included in their professional curriculum and adhered to in practice, have been studied empirically and summarized in a set of "guidelines" for genetic counselling, fetal diagnosis and genetic screening. The core principles are the autonomy of the individual or the couple, their right to adequate and complete information, and maintenance of the highest standards of confidentiality. It follows that the choices to be made are the responsibility of the individual or the couple, and that genetic counselling should be provided in a non-directive manner. Non-directive counselling does not mean simply giving people the facts and letting them make up their own mind. It is a special skill, requiring a special personality and appropriate training. It involves supporting in decision-making those who are being counselled, by helping them to understand their unique medical, social and moral situation.

The application of new genetic technology can create new ethical problems. For instance, whenever a DNA probe or a technique is developed, it is an important medical responsibility to make it available for research. This is sometimes impeded by non-scientific considerations. Because knowledge of the results of predictive tests for severe monogenic disorders may have serious consequences for individuals, the tests need to be highly reliable. An example is the possibility of predictive tests at the DNA level for the gene for Huntington's disease. As a test becomes reliable, it will be possible in principle for individuals in the family of a patient to avoid

9

transmitting the gene to future generations. However, for many, the price will be definite knowledge that they carry the gene for this severe disease, and there is justifiable worry that this could lead to a tragic consequence, such as suicide. There is no easy way out. Not making the test available means that more people, namely the spouses and offspring of the patients, may be subjected to anxiety. It seems reasonable that the spouses of people at high risk for such diseases should share their partners' knowledge, but the responsibility for deciding whether they should or not must rest with the partner at risk. The best way to find out how to use new diagnostic tests is to discuss them with the affected families, as the final responsibility for decision-making rests with them.

There is a possibility that DNA technology will ultimately permit the determination of many normal, as well as pathological, inherited characteristics, in the fetus as well as in the adult. There is a need for guidance on the usefulness of offering tests, especially in cases of genetic variation to predisposition, or resistance, to common disorders. In the majority of cases the predisposing genes do not necessarily themselves lead to disease but, in concert with other genes, can increase susceptibility to disease. Society has no established tradition for handling information on the normal genes of individuals, similar to that for communicating medical information on previous or current diseases to insurance companies and other consumers of medical information. Every measure should be taken to ensure that individuals can undergo testing for such characteristics without fearing that the results might be used to their disadvantage — for example, to limit their choice of employment or to weight their life insurance premiums. To use this information in such a way would be unfair, since tests are available for only a few risks; and testing is essentially random, as regards availability at a given time and in a given place. As our ability to have knowledge of such normal genes grows, it is increasingly important to establish rules of confidentiality. In some countries this could require new legislation.

It is clear that new possibilities for mass genetic screening will arise from human genome mapping procedures. These could well be realized in the near future. Compulsory screening cannot be recommended, since it would interfere with people's autonomy and might cause anxiety. Predictive genetic screening offered on a totally voluntary basis, and followed by adequate counselling, appears to be the only acceptable option.

However, unless precautions are taken, even voluntary screening could lead to unfavourable consequences for the individual. It could be damaging if it were to become known that a person had a greatly increased risk of contracting a serious disorder that could lead to early incapacitation or death, or to mental instability. Accordingly, the data obtained from genetic testing must be strictly protected. Employers, prospective employers, life insurance companies, pension funds, educational institutions, the military, and other consumers of health data should be denied access to the results of predictive genetic tests on individuals. They should be prohibited from requiring such testing or asking whether predictive tests have been performed.

In spite of recent accelerated progress in molecular genetics, gene therapy is not yet possible for genetic diseases. If such therapy becomes possible, it will raise new ethical issues.

CIOMS proposes a dialogue ranging from the presentation of philosophical points of view to practical recommendations, requiring the considered opinions, not only of physicians and researchers, but also of ethicists, lawyers, legislators, theologians and, in particular, the potential consumers of services.

From the agenda for this Conference it is clear that many important questions will be considered during the next five days — the current state of knowledge, the prediction of future trends, and possibilities of implementing in medical practice the achievements made. In fulfilling its constitutional role as the directing and coordinating authority on international health work, WHO is willing to develop its already close cooperation with intergovernmental organizations and agencies, nongovernmental organizations and individual experts in striving towards achievement of the goal of health for all by the year 2000.

I should like to close by wishing you every success in your deliberations during this Conference. I look forward to hearing the outcome with great interest and anticipation.

F. Mayor
Director-General, Unesco*

Mr. Chairman, Ladies and Gentlemen.

I should like to congratulate CIOMS on organizing this Conference and thereby providing a forum where scientists, doctors, parliamentarians, health administrators, ethicists and other lay people can discuss and exchange views on different aspects of research on the human genome and its applications.

Indeed, the interdependence of developments in the world is felt more acutely today than ever before. The human genome project, in particular, which will lead to breakthroughs in the most intimate knowledge of the biology of human beings, requires true international co-operation and an unrestricted exchange of information, firstly because of the returns this can have with relation to research and application and, secondly, because of the need for international analysis of the societal, ethical, and even legal implications this project could engender. Thus, Unesco can only highly commend a conference such as the one taking place this week.

Normally I would be with you today but other commitments have precluded this. I truly regret this both because of the many interesting discussions which the Conference will surely give rise to and because of Unesco's keen interest in matters pertaining to the human genome.

A number of you present are aware that Unesco has recently set up a programme on this topic. This was only appropriate in view of the Organization's vast experience in international cooperation both directly with its Member States and with non-governmental organizations. Moreover, it should be recalled that within the U.N. family Unesco is the only agency handling the basic sciences.

This initiative was therefore fully supported by our Member States and we have already taken some concrete actions to support and promote international cooperation on the human genome. Examples are the Workshop on International Cooperation for the Human Genome Project held at Valencia, Spain, in October 1988; the COGENE/Unesco/EEC/FEBS/ IUB Symposium on Human Genome Research: Strategies and Priorities, which took place at Unesco's Headquarters in January this year; the training course in Techniques in Human Genome Research, and the Symposium and Workshop on Molecular Genetics and the Human Genome Project: Perspectives for Latin America, which were, both, held in Santiago, Chile, last month.

In recommendations submitted to Unesco by scientists who participated in two consultative meeting we held concerning our recently-created pro-

---

* Read by the Conference Chairman in the absence of Professor Mayor.

gramme, it was stressed that one of Unesco's roles in the human genome project will be to serve as a forum for discussion of the social and ethical issues arising from the application of research results. Accordingly, we are now extending our support to the Second Workshop on International Cooperation for the Human Genome Project: Ethics, to take place at Valencia, Spain from 12 to 14 November 1990.

You will appreciate from this that the issues to be discussed today and during the forthcoming days are of prime interest to Unesco. I therefore wish you all a highly successful meeting and I look forward to learning about the results of your deliberations. Thank you.

Z. Bankowski
Secretary-General, Council for International Organizations
of Medical Sciences.

Mr. Chairman, Ladies and Gentlemen.

Many, perhaps most, health policy decisions raise ethical questions. Different national, cultural and religious traditions yield different ethical value systems, and their interactions with health policymaking will therefore vary from country to country. With the aim of strengthening national capacities for addressing and making decisions about the ethical and human-value issues involved in health policymaking, CIOMS initiated, five years ago, a transcultural programme entitled "Health Policy, Ethics and Human Values — An International Dialogue".

CIOMS conferences are aimed at creating international, interdisciplinary and transcultural forums to discuss the implications of scientific and technological progress from social, economic, ethical and legal points of view.

The nature and scope of research into the human genome and the potential of its application in medicine have raised a host of ethical questions, and it is timely to address them now.

The value of the human genome project and its immense potential for good cannot be overemphasized but, at the same time, there is an urgent need to assure that the outcome of this formidable scientific undertaking will be used for the good of individuals and communities and of future generations.

Society must be assured that the exploration of the human genome and its biomedical applications will not degrade or impair human dignity. Continuous dialogue on the related ethical issues is necessary between the scientists engaged in research into the human genome, the medical profession, ethicists, policymakers and society at large. The public must be kept continuously informed of the results of the study of the human genome so that they will understand and accept this extraordinary advance in biology and medicine. Only thus can we ensure that the fruits of this project will benefit humanity and that risks will be minimized and well controlled.

Clearly, the best guarantee of the responsible use of molecular genetics in medicine, and the best protection against its misuse, is a well-educated, well-informed public. Molecular genetics needs to be included in education, from primary school onward, if informed, rational, public participation is to be assured in discussions and decisions about its use.

The participation in our Conference of the Director-General of the World Health Organization, Dr. Hiroshi Nakajima, and the co-sponsorship of its proceedings by WHO and Unesco, are the best guarantee, for all of us, that necessary action will be taken at the global level to maximize the benefits and minimize the risks associated with the application of advances in molecular genetics.

A continuing international and transcultural dialogue on the application of molecular genetics in medicine is an essential means of safeguarding the fundamental principle of medical ethics: *do good and not harm.*

KEYNOTE ADDRESS

James Wyngaarden[*]

The broad sponsorship of this conference, and the participation of scientists, ethicists and policy-makers from many countries, are significant in underscoring that, just as science knows no national boundaries, so too the ethical and social issues that we shall address are international in scope. The issues are germane to developed and developing countries alike. Particularly in this historic year of the re-emergence of personal freedoms and human rights in many political areas where they have been until recently difficult to express, it is imperative that science too be emancipating and that it not be used inappropriately to restrict personal liberty or opportunity. In these complex issues it is important that we think clearly and think globally.

The current efforts directed at constructing an increasingly detailed map of the human genome, and eventually obtaining a complete sequence, are being undertaken in the conviction that the information contained in the human genome is worth knowing, and that the information in the genome of man that is not present in the genome of a mouse, for example, will tell us much of interest concerning the nature of man. In addition, there is a great expectation that powerful new insights will emerge concerning not only those several thousand diseases that are inherited as Mendelian disorders, but also a possibly equally large number of common diseases or susceptibilities in which genetic components are believed to play a part. It is also a tenet of this faith that with a deepened understanding of the nature of human afflictions will come new insights for prevention and therapy. Walter Bodmer even predicts that the pharmaceutical industry of the future will rest upon the association of certain sequences with certain disease susceptibilities, and new insights in therapeutic approaches to rheumatoid arthritis, diabetes, Alzheimer's diseases and other common disorders that are conditioned by genetic factors. Understanding cancer as a genetic process and finding new ways of preventing and treating the disease do not seem to raise any particular ethical issues, except perhaps the imperative to proceed apace in order to achieve these insights as soon as possible.

The objective of mapping and sequencing the human genome has become a part of national policy in the United States of America as well as in several European countries and Japan. A non-governmental association known as the Human Genome Organization, or HUGO, has been established to help coordinate, integrate and plan these activities on an international scale. An office for the Americas is located in Bethesda, Maryland and

---

[*] Foreign Secretary, National Academy of Sciences, Washington D.C., USA.

one for Europe in London, and discussions are under way concerning the location of an office for the Asian countries in Japan. The objectives of HUGO are to facilitate international collaboration in the understanding of the human genome and to help facilitate the worldwide research effort so that all scientists who wish to do so may participate, and that the mapping and sequencing information is freely available. Much attention is being given to the logical progression of this massive and lengthy effort, and to ensuring efficiency, economies of scale and minimal redundancy.

The enhanced ability to deal with genetic disease and genetic components of common disease that will progressively result from the Human Genome Project will expand our ability to detect an ever larger number of these conditions *in utero*, or even before, and will expand opportunities for decisions about reproduction. The ability to predict at an early age the onset of disease at a later age already exists in a few situations, of which Huntington's disease has become the prototypical example. Our ability to make genetic diagnoses, or to predict susceptibility to a late-onset disease, has moved far ahead of our ability to intervene therapeutically. Many questions have already been raised. Should prenatal diagnosis be performed for non-lethal conditions, or for conditions of variable expression, which sometimes are rather mild? Should newborn screening be done for conditions for which there is no treatment? Should early detection be attempted for conditions of late or uncertain appearance? If screening or diagnostic tests are performed, who should be told the results? When should they be told? If a subject is told, who else should or may know? Should the spouse, relatives, the employer, or the insurer have access to such information? Some insurance companies have held that they should have access to such information if the applicants or policyholders have access to it. If testing is done, should it be voluntary or mandatory? Almost everyone agrees that a high order of confidentiality should be preserved about genetic test results, but how can confidentiality be assured? We now have the ability not only to store information extracted from DNA, but also to bank the basic genetic material itself. Yet there are few guidelines to aid DNA banks in handling such information.

Early controversies about attempts to introduce carrier detection programs for sickle cell anemia and Tay Sachs disease led to a consensus that programmes should, at a minimum, fulfil four major criteria:

1. The condition must be serious;
2. Diagnostic tests must be accurate;
3. Therapy or other meaningful intervention must be available; and
4. Screening goals must be achievable at reasonable costs, or at an acceptable cost-benefit ratio.

We may wish to consider whether these or other criteria are generally applicable to carrier detection.

At his first press conference, on his appointment as Associate Director for Human Genome Research at the National Institutes of Health, Nobelprizewinner James Watson stated that from the inception of the programme some funds would be devoted to the study of ethical, legal and social issues related to mapping and sequencing the human genome. At its first meeting, in September 1989, the Joint National Institutes of Health — Department of Energy Working Group on Ethical, Legal and Social Issues identified nine topics of relevance to the genome project:

1. The fair use of genetic-test information in such areas as insurance, employment, criminal justice, education, adoption and the military;
2. The impact of genetic information on individuals;
3. Personal privacy and confidentiality of genetic information;
4. The impact that the dramatic increase in human genetic information will have on genetic counselling and the delivery of genetic services;
5. The influence of genetic information and new technologies on reproductive decisions;
6. Issues raised by the introduction of new genetic information and technologies into mainstream medical practice;
7. The historical analysis of the use and misuse of genetic information and technologies;
8. Issues raised regarding the commercialization of research results; and
9. Conceptual and philosophical questions related to human genetics.

The National Institutes of Health and the Department of Energy issued programme announcements in January and March 1990, respectively, requesting research proposals on these topics. It is hoped that many ethicists, social scientists, humanity scholars and legal analysts will join with and assist their medical and scientific colleagues in addressing these issues. This conference will certainly make an important contribution to the inquiry and the debate.

I am pleased that Dr Bankowski, and the other members of the programme committee who crafted the agenda of this meeting, included not only human genome mapping and genetic screening, but also genetic therapy.

Discussions of the ethics of gene therapy in human beings have occurred over many years. Most participants have concluded that somatic-cell gene therapy, conducted for the sole purpose of medically correcting a severe genetic disorder, involves no new ethical principles comparable to, for example, those encountered in organ transplantation for a similar purpose. Somatic-cell gene therapy in such situations would be ethically acceptable if carried out under the same strict criteria that apply to other new experimental medical procedures. In the instance of somatic-cell gene therapy, three general criteria have been proposed for animal studies before a clinical trial is conducted in human subjects. These are:

1. That the new gene can be put into the correct target cells and will remain there long enough to be effective;
2. That the new gene will be expressed in cells at an appropriate level; and
3. That the new gene will not harm the cell, or by extension, the host.

These criteria are very similar to those applicable to the use of any new drug, therapeutic procedure or surgical operation. They simply state that the new treatment should get to the area of the disease, correct it and do more good than harm. One particularly important aspect of gene therapy experiments, which does not apply to new drugs or therapeutic procedures in general, is the use of retrovirus or other viral vectors and the degree of safety that can be assured concerning their use.

The first approved experiment involving gene transfer to somatic cells of a human recipient occurred at the National Institutes of Health about one and a half years ago. It was not gene therapy, but rather the transfer of a bacterial marker gene for neomycin-resistance to activated tumor-infiltrating lymphocytes (TIL) or TIL cells that were subsequently returned to the donor as experimental cancer therapy. The proposed protocol went through eight levels of committee review at the NIH and the Food and Drug Administration in order to satisfy biosafety and ethical questions. The final review was by the Recombinant DNA Advisory Committee in public session. Only after all of these hurdles had been surmounted did I, as Director of NIH at that time, give my approval to the scientists to proceed. The initial approval was for ten subjects, all with potentially terminal cancer. Of course, informed consent was a requirement.

The procedure has now been performed on seven or eight patients without untoward effects that could be ascribed to the marker gene. The presence of the marker gene has contributed important information on the target sites to which TIL cells have gone, and on the biological turnover of the TIL cells. Permission has now been given by the Recombinant DNA Advisory Committee and the Acting Director of NIH to extend the series beyond ten patients.

Two additional proposed gene-marker experiments are currently under review at the NIH, one from St. Jude Hospital in Memphis, Tennessee, and the other from the University of Wisconsin Medical Center in Madison, Wisconsin. Both involve marking bone-marrow transplantation cells used in treatment of leukemia patients. The first protocol concerns children, the second adults.

There is now another protocol under review, one for what could be the first approved attempt at *bona fide* human gene therapy involving replacement of a missing or defective gene: the possible transfer of the gene for adenosine deaminase into patients with ADA deficiency and severe combined immunodeficiency disease. The protocol has been approved by the biosafety committees of both the National Heart, Lung and Blood Institute and the National Cancer Institute. Next it will be reviewed by the Human Gene Therapy subcommittee of the Recombinant DNA Advisory Committee as well as by the U.S. Food and Drug Administration.

Another protocol, for gene therapy of a quite different type, is in the early stages of internal review at NIH; it proposes the addition to the marked TIL cells of a gene for tumour necrosis factor. If this protocol passes the sequential review hurdles, it will represent the first attempt at cancer therapy with cells that have been engineered to provide a tumour toxin.

I have several times used the term "approved protocol", mindful of the fact that, regrettably, there had been earlier unsanctioned and ill-advised attempts at human gene therapy, which reflected very badly on all of us. These were the experiments of Rogers and of Cline in the United States, attempts to correct arginase and sickle-cell defects by the transfer of genetic material.

Whereas the possibility of human gene therapy involving somatic cells has moved forward cautiously with respect to both experimental and ethical issues, the same cannot be said of gene therapy involving reproductive cells. I know of no proposals for experiments involving heritable traits in man, nor in the present state of knowledge would it be appropriate to propose such experiments, in my view.

Much has been written in newspapers, magazines and books about public apprehension about tampering with hereditary traits in man. Although many of the project scenarios are merely inflammatory science fiction, far beyond the capabilities of rational science — for example, transfer of genes for intelligence — it is nevertheless beneficial that issues raised by germ-cell therapy be clearly debated well in advance of any such proposal. For example, will interventions involving germ cells be ethically acceptable in any circumstances? If so, what criteria, what principles, should apply? Germ-cell therapy may indeed raise larger issues than those raised by human-gene transfer: it may include genetic pharmacology, for example. Some scientists believe that certain genetic defects, such as single-base substitutions or limited deletions or insertions, are theoretically correctable by the clever use of antisense-DNA or -RNA molecules plus normal cellular enzymatic corrective mechanisms. If such an approach should at some time prove capable of correcting the sickle-cell defect or Tay Sachs defect in the germ line, what would the ethical reaction be? What sort of feasibility and safety criteria would need to be met?

This conference will have an opportunity to address certain of these issues, particularly in the satellite symposia on biotechnology and human genetic diseases, where sessions on antisense cDNA and on neurotropic gene therapy for Lesch-Nyhan patients are included.

In closing, let me thank Dr Yagi and all of the members of the organizing committee for the extraordinary contribution that they have made in arranging all the myriad details of this conference. We look forward to several days of stimulating and profitable discussion on subjects of vital importance. We also look forward to the establishment of closer ties with our Japanese colleagues, and to the opportunity to visit and enjoy your beautiful and remarkable country. For all of this we are indeed grateful.

HUMAN GENOME MAPPING: APPLICATION TO GENETIC DISEASE

R.G. Worton*

The map of the human genome is the subject of world-wide interest, on the part of not only geneticists but also physicians, biologists, chemists, physicists and even ethicists and lawyers. This is a substantial change from only five years ago when genetic mapping, and the resulting maps, were of interest to only a handful of the most dedicated geneticists. They knew something that I did not know: that the map would form the underpinnings for the most grandiose undertaking in the history of biology — the international effort to map and sequence the entire human genome. Little did I know in 1972, when I refused to become involved in the mapping of human genes (I thought it too boring!), that a decade later I would embark on a project to identify and clone the gene responsible for one of the most common and most devastating of genetic diseases, utilizing a strategy based upon that very map. The success of this project and others like it has convinced me and others that the key to the understanding of the 4,000 or so genetic diseases lies in the identification of the genes that are mutated in the affected individuals. By extension, the key to the understanding of all human biology lies in the identification of the 50,000-100,000 genes encoded in the 22 autosomes and two sex chromosomes of the species. This is the heart of the human genome project. The mapping and the sequencing are the tools to get there.

Gene mapping technology

Somatic cell hybrids and hybrid panels. Human gene mapping dates to 1968 when the technology emerged to create somatic-cell hybrids by fusing human cells to rodent cells. The technology is based on the fact that human chromosomes are lost from hybrid cells, quickly at first, and then more slowly, until at the end of a few weeks or months of culture a typical hybrid-cell line contains only a few human chromosomes. This 'segregation' of human chromosomes from hybrid cells made it possible to correlate the presence of a human gene product in the hybrid-cell line with the presence of the human chromosome carrying the gene encoding that product. The basic procedure was to create a set of independent human/rodent hybrids and test each for the presence or absence of human chromosomes and human genetic markers. Once these 'panels' of hybrids were established and well characterized in terms of their human chromosome content, it became relatively easy to check them for the presence or absence of new genetic markers as these became available.

---

* Geneticist-in-Chief, Hospital for Sick Children, and Professor, Department of Molecular and Medical Genetics, University of Toronto, Toronto, Canada.

In the 1970s the major effort centred on the mapping of genes encoding human enzymes, as the human enzyme activity could usually be separated from the rodent activity by electrophoresis, thereby allowing hybrid cells to be tested for the presence or absence of the human gene product. In the 1980s the emphasis turned to the mapping of DNA fragments by means of Southern blot technique to determine the presence or absence of a particular DNA sequence in a hybrid cell line. The segment of DNA to be mapped was often derived from a cloned gene but was sometimes an anonymous segment of DNA from a library of human genomic DNA. If the gene or anonymous segment hybridized to a region of the genome at the site of a restriction fragment length polymorphism (RFLP) or other polymorphic marker sequence, then the same DNA fragment could also be used as a probe to follow the segregation of the polymorphic marker in families (see below).

The somatic-cell hybrid technology, initially used to map markers to individual chromosomes, has become equally valuable for the regional mapping of markers within chromosomes, taking advantage of the huge number of natural chromosome rearrangements that exist in the human population. Thus, hybrid cell lines carrying parts of human chromosomes, derived from individuals with well-characterized translocations and deletions, form mapping panels that may be used to localize markers regionally. For some chromosomes these panels are quite extensive, dividing a chromosome into 20 or more mapping intervals.

Radiation hybrid mapping. A method for dividing a chromosome into even smaller intervals, described initially by Goss and Harris, has been developed into a sensitive mapping technique by Cox.[1] Essentially, a hybrid cell line carring one human chromosome is given a lethal dose of radiation, smashing the chromosomes into tiny fragments. The fragments are then "rescued" by fusing the irradiated cells with new rodent cells, and the secondary hybrid cell lines are found to contain one or a few small pieces of the human chromosome integrated into the rodent chromosomes. While the human chromosomal fragments are not large enough for their origin to be recognizable under the microscope, the 'radiation hybrid' and the chromosomal segment that it carries can be characterized according to its content of previously mapped markers. Markers found frequently together in the same hybrid are assumed to be close together in the chromosome.

In situ hybridization. An alternative mapping technique depends on the direct hybridization of DNA segments to spreads of metaphase chromosomes on microscope slides. The hybridizing DNA is radioactively labelled and detected on the slide by overlaying with photographic emulsion. The grains in the emulsion depict the location of the DNA probe on the chromosome. More recently, the technology has developed to attach biotin to the hybridizing DNA, permitting its detection with fluorescent compounds attached to avidin, through the very high affinity bond in an avidin-biotin complex.[2]

Linkage analysis in families. The mapping of DNA segments on chromosomes as described above provides physical mapping information, and linkage studies provide genetic mapping information. The latter measures genetic distance between two markers, where the unit of measurement is related to the frequency with which two markers segregate together from one generation to the next in a family. Markers on different chromosomes will segregate together by chance 50% of the time. This is true also of markers on the same chromosome that are far apart, since the probability of a recombination event between the markers, switching the phase of the markers, is close to one. Markers that segregate together significantly more frequently than 50% are said to be 'linked', and the 'linkage distance' is measured in centi-Morgans (cM): one cM is the chromosomal distance over which the frequency of recombination is 1%. Averaged over the genome, a genetic distance of one cM corresponds to a physical distance of about 1,000 kilobases (kb) or one megabase (Mb). In practice, however, recombination frequency per unit length of DNA is not constant over different regions of the chromosome, so the relationship between genetic distance and physical distance varies according to location in the genome.

Genetic mapping, based on the segregation of genetic markers in families, lagged behind physical mapping until Botstein et al.[3] proposed the use of human genetic variation in the form of restriction fragment length polymorphisms (RFLP) as genetic markers to develop a detailed genetic map. Since 1980 the genetic map has been developed quickly, with the mapping of numerous genes as well as anonymous segments of DNA.

Table 1. Human Gene Mapping Workshops Summary

Year	Site	Genes & Markers Mapped	Cloned Genes	Anonymous DNA segments Mapped	DNA Polymorphisms	TOTAL
1973	New Haven	64	—	—	—	64
1974	Rotterdam	125	—	—	—	125
1977	Winnipeg	176	—	—	—	176
1979	Edinburgh	230	—	—	—	230
1981	Oslo	319	16	35	18	354
1983	Los Angeles	620	104	215	95	835
1985	Helsinki	831	249	559	245	1479
1987	Paris	1301	610	2057	977	3462
1989	New Haven	1740	945	3417	1495	5157

Note: The data in column 4 (cloned genes) are a subset of the data in column 3 (genes and markers mapped); the data in column 6 (DNA polymorphisms) are a subset of the data in column 5 (anonymous DNA segments mapped).

Mapping of genes and genetic markers

The human gene mapping conferences. Since data began to accumulate on the human gene map, workshops have been held at approximately two-year intervals. Known as the Human Gene Mapping Conferences (HGM 1-10), these ten conferences have fulfilled the role of reviewing the literature and examining new data to update the map. The proceedings have been published regularly in *Cytogenetics and Cell Genetics.* Table 1 provides a summary of the number of markers, including anonymous DNA segments and genes, that have been mapped at each of the conferences. At HGM 1 in New Haven 64 markers, including 25 known genes, were included in the map. Two of these assignments were subsequently proven to be wrong, and at later conferences map results were considered provisional until an independent confirmation became available. During the 1970s there were no cloned genes, and most of the genetic markers were genes encoding enzymes detectable in somatic cell hybrids, or well-known genetic traits such as blood groups. Later, in the 1980s, this source of genetic markers became saturated, and cloned genes as well as randomly cloned segments of DNA became available for mapping. Disease genes began to be mapped by the segregation in families of the disease phenotype with other mapped polymorphic markers, and this activity increased as the number of DNA polymorphisms expanded. Finally, as disease genes themselves began to be cloned, these also were added to the growing map. The methods of cloning disease genes are discussed briefly below.

Cloning disease genes with known protein products

Cloning genes with abundant mRNA. In the early days of human molecular genetics (i.e. 1978), the first disease genes to be cloned and characterized were those for which the protein product was already known, and for which messenger RNA was readily available. Thus, mRNA from the α and β globin genes was prepared from reticulocytes, where it exists in great abundance, and complementary DNA (cDNA) was prepared by reverse transcription and cloned into a bacterial plasmid. The cDNA clones were then used as hybridization probes to identify clones of chromosomal DNA fragments that contained the gene that gave rise to the message. The finding that the chromosomal-derived (genomic) DNA differed from the cDNA copies of the message by the inclusion of genomic sequences that were removed in the generation of the message led to our current undestanding of eukaryotic genes as being made up of exons (coding regions) and introns or intervening sequences.

Cloning genes with less abundant mRNA. Next followed a period (1980-85) when genes encoding many known proteins were cloned and characterized, even though their mRNA was not abundant in the cell. Essentially the bacterial colony or phage plaque carrying the gene of interest had to be indentified among hundreds of thousands of organisms containing a 'library' of DNA fragments. Two approaches were used. In some cases a protein was purified and a small amount of amino-acid sequence was obtained. This permitted the determination of the corresponding nucleotide sequence of the mRNA, accounting for redundancies in the code. This sequence of bases was then synthesized and the synthetic 'oligonucleotide' was used as a probe to identify bacterial or phage clones carrying the gene of interest. In other cases an antibody was raised against the purified protein and the specific antibody was used to identify the clones that were producing the relevant protein from human DNA ligated into an appropriate "expression" vector. This scheme is depicted on the left side of Figure 1.

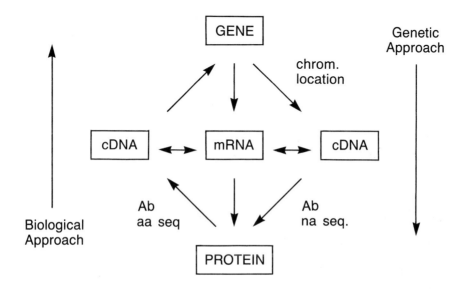

Figure 1. The 'biological approach' and the 'genetic approach' to obtaining cDNA clones.

The figure revolves around the central paradigm that the gene is transcribed into mRNA, which is translated into protein. In the test-tube, mRNA may be readily copied by a "reverse transcriptase" into cDNA (horizontal arrows). In the biological approach to the study of genetic disease, the basic defect is first identified as a faulty protein and the amino-acid sequence of the protein or an antibody against the protein is used to identify the correct cDNA clone from a cDNA library. The gene itself is identified from a genomic library, by the use of the cDNA as a hybridization probe. In the genetic approach the reverse path is followed, the first step being the mapping of the gene to a chromosome and then its mapping to a specific region of the chromosome. DNA from the region of interest is cloned and used as a hybridization probe to identify, from a cDNA library, the correct cDNA. The cDNA can then be sequenced and also transcribed and translated into protein to produce all or part of the protein product. Antibodies raised against this product are then used to identify the native protein that is defective in the disease.

Cloning disease genes from knowledge of their location on the map

Finding a gene by screening a cDNA library with an oligonucleotide probe or with an antibody against the protein is feasible only when the protein product of the gene is known. Among the 400 or so genetic diseases, only a small proportion are caused by defects in known proteins. These include, for example, the hemoglobinopathies (globin genes), the hemophilias (Factor VIII, Factor IX genes) and several of the inborn errors of metabolism of which the enzyme defect is known.

For the remaining diseases, many of them common and lethal diseases, the basic defect, and therefore the identity of the faulty protein, is completely unknown. By about 1982 new technologies and new approaches opened up the possibility of first identifying the responsible gene and from that determining the nature and function of the protein product. The problem was that of identifying a gene of perhaps 3-30 kilobases (kb) in size from a genome containing 3 million kb of DNA. Rather than using information about the gene product as a starting point, the new approach used information about the position of the gene in the chromosome. Briefly, DNA from the chromosomal region of interest is cloned, and each piece is examined in detail in order to identify all genes in the region. These are then tested, one at a time, to find the one that is altered (mutated) in individuals with the disease. The approach has been termed "reverse genetics" to distinguish it from the more traditional approach of cloning from knowledge of the protein. This is an unfortunate term since the approach is truly genetic in nature, depending for its success on the application of many of the principles and techniques of genetics. The approach is depicted on the right hand side of Figure 1.

The genetic approach to disease gene cloning. Figure 2 illustrates the genetic approach in more detail, showing the paths that have been taken to arrive at a cDNA clone from a disease gene. For a handful of genetic diseases, chromosomal rearrangements have been found in association with a specific genetic disease (left side of Figure 2). In the case of Duchenne muscular dystrophy (DMD), for example, a few atypical affected individuals were found to have a translocation between the short arm of the X chromosome at band Xp21 and one of the 22 autosomes. The translocation causes the disease by disrupting the DMD gene at Xp21, and in our laboratory the cloning of DNA from the translocation junction of one such individual has enabled us to identify sequences from the DMD gene. Other DMD patients were found with deletion of the X chromosome around band Xp21, and by a clever subtraction hybridization technique Kunkel and colleagues were able to clone DNA from the deleted region and to identify sequences from the DMD gene (reviewed in[4,5]).

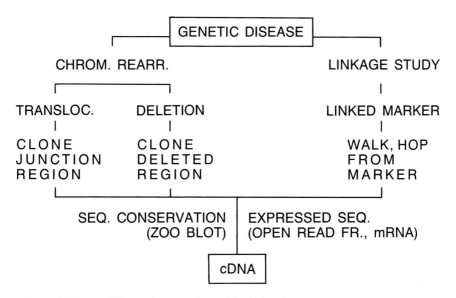

Figure 2. The possible genetic approaches to identifying the cDNA encoded by a gene whose mutation is responsible for a genetic disease.

For many genetic diseases there have been no chromosomal rearrangements to point the way to the location of the gene on one of the chromosomes. Cystic fibrosis (CF) is an example. Without this, it was necessary to find the gene through an arduous family-linkage study with multiple genetic markers — the right-hand path in Figure 2. The procedure was to take anonymous DNA segments or genes that recognized polymorphic markers and use each as a probe to trace the markers in CF families. If a marker failed to segregate with the CF phenotype in the family it was rejected and another one was tried. After about three person-years of effort, and exclusion of about one third of the genome, a linked marker was found.[6] It mapped to the long arm of chromosome 7, focusing attention on this chromosome. Quickly, new markers were identified that were closer to the CF gene, the closest being about 1 cM (1 million base pairs) from the gene. From then a second arduous phase ensued in which chromosome walking and hopping was used to close the distance between the nearest marker and the gene. Eventually, the combined effort of three laboratories led to the identification of DNA sequences from the region of the CF gene.[7,8]

By all three paths illustrated in Figure 2 the clonal isolation of DNA from the chromosomal region around the gene is only the first step. To find the expressed sequences (exons) of the disease gene requires further analysis. In most cases the next step is to search for small regions that are conserved through evolution by demonstrating hybridization of the human DNA to DNA from a variety of organisms in a cross-species Southern blot or "zoo blot". Conserved sequences are much more likely to come

27

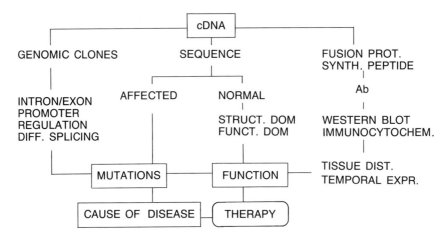

Figure 3. The possible routes to follow once a cDNA from a genetic disease gene is identified.

from a genetic coding region (exon) of a gene, and, once identified, the conserved sequence is then used as a probe to identify an homologous cDNA clone from a cDNA library. The cDNA clone then becomes a *candidate gene* for the disease and must be tested for the presence of mutations in affected individuals.

Charaterization of candidate cDNA clones. Figure 3 presents the course of the research after a cDNA clone has been obtained, experiments being designed to determine the nature of mutations in the gene and to identify and characterize the protein product of the gene. One of the first courses to follow is to sequence cDNA clones from a normal individual and from an individual affected with the disease — the middle path in Figure 3. Any sequence difference between the two clones potentially represent the genetic defect in the patient. Sequence comparison with a number of normal individuals and affected individuals will determine whether the difference represents the mutation responsible for the disease.

The sequence of the cDNA also provides the translated amino-acid sequence of the protein. Sequence comparison with other known proteins may reveal sequence homologies that give clues to the structural and functional domains of the unknown protein. This type of information permits intelligent guesses about the function of the unknown protein.

Following the left-hand path in Figure 3 it is also possible to isolate genomic sequences from a library of chromosomal DNA and characterize these in terms of their intron/exon boundaries and their genetic regulatory sequences. Again, by comparing normal with affected individuals, it is often possible to detect sequence changes that alter the splicing pattern, thereby giving rise to altered mRNA and faulty protein. Analysis of the promoter region of the gene may also reveal alterations that result in reduced transcription and, therefore, inadequate amounts of the protein product.

Following the right-hand path in Figure 3, it is possible to use a portion of a cDNA clone, joined to a bacterial gene, to create a genetic construct that, when put into bacterial cells, will give rise to a bacterial/human fusion protein. This fusion protein contains amino-acid sequences from the disease-gene product which, when injected into rabbits, will produce antibody directed against the protein of interest. Alternatively, short synthetic peptides can be prepared, based on the derived amino-acid sequence; these can be conjugated to larger proteins and injected into rabbits to make antibody directed against the peptide. Both the fusion-protein antibodies and the peptide antibodies provide reagents to detect the protein of interest by western blot analysis and by immunological staining of tissue sections. Analysis of a variety of tissues during the course of development gives information about the tissue distribution and the spatial and temporal expression of the protein. This information, when coupled with the information on the structural and functional domains of the protein, may provide fairly detailed knowledge about the biological function of the protein.

The ultimate goal in the study of genetic disease is an effective therapy based upon the knowledge of the basic defect in the disease. As can be seen from the bottom of Figure 3, it should be possible to arrive at a therapeutic approach by following the three paths, leading to a detailed understanding of the gene and its protein product.

Application to the DMD and the CF genes. Application of this scheme to the gene for Duchenne muscular dystrophy led first to the discovery that the cDNA was large (14 kb) and that on Southern blots it identified chromosomal segments spanning 2,300 kb of the X chromosome. This identified the gene as the largest ever discovered. Analysis of patient DNA revealed that 60% of patients had deletions of one or more exons from the DMD gene, and another 6% duplications of exons in the gene. No such deletions were found in the DNA from normal individuals, confirming that the cloned gene was indeed the one responsible for the disease. Generation of antibodies by the injection into rabbits of fusion proteins and synthetic peptides resulted in the identification of dystrophin, a 427kD protein found primarily in cardiac and skeletal muscle. Immunocytochemical tests on muscle sections revealed that the immunoreactive protein was localized to the sarcolemmal membrane of the muscle fibre in normal individuals. The protein was not present in the muscle of affected individuals. Sequencing of the cDNA revealed sequence homologies with two different cytoskeletal proteins, indicating a possible cytoskeletal role for dystrophin. The presumed function of dystrophin, therefore, is to provide a structural component on the inner surface of the cell membrane; when this molecule is missing, the membrane is weakened and susceptible to contraction-induced tearing. This would explain the degeneration of the muscle characteristic of the disease.[4,5] This new understanding suggests that a rational therapy for the disease must involve the replacement of dystrophin in muscle tissue. To this end, myoblast transplantation is being considered, as injected myoblasts fuse with muscle tissue and provide dystrophin-competent nuclei for the production of dystrophin in the muscle. This approach has met with success in a mouse model of the disease,[9,10] and pilot projects are beginning in affected boys.

In the case of the CF gene, sequencing of the cDNA led to the discovery of a 3 bp deletion found in 70% of all the chromosomes carried by CF children.[7,8] This deletion has never been found in normal (non-CF) chromosomes. The sequence analysis further revealed that the translated amino-acid sequence contains several characteristic transmembrane domains. Also, the overall structure of the protein, as determined from its amino-acid sequence, bears strong similarity to a class of transmembrane proteins responsible for multi-drug resistance. The 3 bp deletion resides in a regulatory region of the protein. This information, coupled with the physiological data suggesting that the defect in CF children is related somehow to the regulation of chloride transport, has resulted in the name *cystic fibrosis transmembrane conductance regulator* (CFTR) being assigned to the protein.[8] Subsequent analysis of additional patient DNA has identified many additional mutations besides the 3 bp deletion that accounts for 70% of the mutant chromosomes. Antibodies are now being prepared against the protein in order to study its tissue distribution and temporal regulation.

Impact of gene cloning on human health

There can be no doubt that cloning of the genes for such diseases as Duchenne muscular dystrophy and cystic fibrosis will greatly affect health care. It is likely that the human genome project over the next ten years will result in the identification and characterization of the disease genes responsible for every major genetic disease. In each case the protein product of the gene will be identified and its structure and function will be determined. Additional genes will be identified whose protein products are involved in cellular growth control; indeed, several genes already cloned are of this type and are therefore involved in cancer susceptibility. Another class of genes that will be more difficult to identify, but will ultimately be isolated and characterized, are those involved in multifactorial diseases in which there is a genetic component. These would include such diseases as diabetes, schizophrenia, Alzheimer's disease and certain cardiac diseases, and many others. For these diseases the genetic defect alters the predisposition to develop the disease rather than causing the disease itself. Thus, detection of a mutant gene redefines a risk for the individual, but it does not follow that the disease will occur.

Carrier identification and prenatal diagnosis. In the case of serious genetic diseases manifest in children, one of the initial benefits of gene cloning will be carrier identification and prenatal diagnosis to prevent the birth of affected children. This is already a reality for the thalassemias, sickle-cell disease and the hemophilias, as well as for many other diseases, including Duchenne muscular dystrophy and cystic fibrosis. At present, carrier detection is limited to families that have already had an affected child, for the information about the defective allele carried in the patient is a necessary part of the carrier identification. This permits the prevention of second and third cases in families, as well as the prevention of affected children in the extended family.

Only in exceptional circumstances does identification of the gene allow for general population screening to detect carriers. This would be the case, for example, for certain alleles of the β globin gene that cause thalassemia with high frequency in specific ethnic populations. In the case of cystic fibrosis, the discovery that a single mutation accounts for 70% of the mutant alleles in affected North American children suggested that a population screen for this allele might be carried out to detect many of the one in 20 Caucasians who carry a CF gene. The practicality of doing this has been reduced in recent months as it has become clear that the remaining 30% of the mutations are caused by a large number of separate mutations, each with a low frequency in the population. In response to pressure from the scientific and medical community for CF population screening, the American Society of Human Genetics convened an *ad hoc* committee to work together with the National Institutes of Health to develop a policy on population screening. In essence, the committee recommended that population screening, if it becomes available, should be voluntary and confidential, that it should be available to all who want it, that informed consent be required, and that laboratory quality assurance begin immediately. It put the onus on health care providers to ensure that adequate education and counselling are available before they offer testing.

The future of genetic testing. Policies developed for CF and other genetic diseases will no doubt have to be revised at regular intervals as new data emerge on the mutations involved in the diseases. Also, technology is constantly improving and becoming more accurate and more rapid, typified by the recent application of the polymerase chain reaction (PCR) as a technological substitute for the time-consuming procedures of cloning and Southern blotting. As PCR reactions become automated and as ^{32}P labelling of DNA is replaced by fluorescent dye labelling, costs per test are expected to drop to the point where large-scale population screening will become practical. At that time the social and ethical issues that have begun to emerge through the study of a few genetic diseases (muscular dystrophy, cystic fibrosis, retinoblastoma, Huntington's desease, etc) will be magnified many-fold. Also contributing to this magnification will be the cloning and characterization of genes that predispose to the development of complex disorders such as diabetes, heart disease, and schizophrenia. This eventuality will make it possible to predict with some accuracy at an early age who will be prone to develop heart disease or neurological deficit. The implications for society are significant as it will become necessary to wrestle with questions concerning the right to know (or not to know), and confidentiality of information, especially with respect to employers and insurance companies.

31

References

1. Cox, D.R. et al. Segregation of the Huntington Disease region of human chromosome 4 in a somatic cell hybrid. *Genomics* 1989; **4:**397-407.
2. Lichter, P. et al. High-resolution mapping of human chromosome 11 by in situ hybridization with cosmid clones. *Science* 1990; **247:**64-68.
3. Botstein, D. et al. Construction of a genetic linkage map in man using restriction fragment length polymorphisms. *Am J Hum Genet* 1980; **32:**314-331.
4. Worton, R.G. & M. Thompson. Genetics of Duchenne muscular dystrophy. *Ann Rev Genet* 1988; **22:**601-629.
5. Monaco, A.P. & L.M. Kunkel. Cloning of the Duchenne/Becker muscular dystrophy locus. In: Harris H, Hirschorn KH, eds. *Adv Hum Genet* 1988; **17:**61-98.
6. Tsui, L-C. et al. Cystic fibrosis locus defined by a genetically linked polymorphic DNA marker. *Science* 1985; **230:**1054-1057.
7. Rommens, J.M. et al. Identification of the cystic fibrosis gene; chromosome walking and jumping. *Science* 1989; **245:**1059-1065.
8. Riordan, J.R. et al. Identification of the cystic fibrosis gene: cloning and characterization of complementary DNA. *Science* 1989; **245:**1066-1073.
9. Karpati, G. et al. Rapid Communication: Dystrophin is expressed in mdx skeletal muscle fibers after normal myoblast implantation. *Am J Path* 1989; **135:**27-32.
10. Partidge, T.A. et al. Conversion of mdx myofibers from dystrophin-negative to positive by injection of normal myoblasts. *Nature* 1989; **337:**176-179.

THE HUMAN GENOME PROJECT IN JAPAN: CURRENT STATUS AND FUTURE DIRECTION

N. Shimizu*

I. The national projects

The human genome efforts in Japan may have stemmed from the "Wada project", which began in 1981 and was aimed at the development of an automated DNA sequencing machine. This was carried out as a project of Riken (The Institute of Physical and Chemical Research) and was supported by the Kagicho (Science and Technology Agency) until 1988. It was directed by Professor Akiyoshi Wada (University of Tokyo), who conceived of the important strategy that "linking of various existing devices into an integrated system is essential for rapid DNA sequencing and data analysis". Professor Wada and his colleagues assembled several components into a semi-automated sequencing system, and he declared that "sequencing of the entire human genome can be accomplished with existing technologies". The plan of performing massive DNA sequencing by using robotics was fascinating to many molecular biologists throughout the world, particularly in the United States. Several private companies were involved in the Wada project.

Despite these pioneering efforts, Japan's official human genome project started only very recently. In July 1989, the Gakujutsu-Shingikai (Council for Science and Education) presented a Kengi (a special report) to the Minister of Education, Science and Culture, emphasizing the importance of establishing the human genome project in Japan. In October 1989 the Ministry formed a committee of six members to plan the Human Genome Programme; it is chaired by Professor Ken-ichi Matsubara (Osaka University) and has been awarded 600 million yen ($4 million) for two years to organize the initial two-year programme and to plan for the next five-year project (1992-1997). The planning committee is backed up by a Task Force and is reviewed by an advisory committee. An appeal for cooperation in human genome analysis was prepared by the planning committee and distributed to several hundred people in various fields to solicit ideas and opinions. In November 1989, an open forum was held at the annual meeting of the Molecular Biology Society to discuss the human genome project. Opinions among 700 participants were generally in favour of the project but there were serious concerns about how to fund it, as was discussed at the Cold Spring Harbor Meeting. In December 1989, the first human genome workshop was held, at which over 100 scientists presented their recent research and over 350 participants discussed the current status and prospects of the project.

* Department of Molecular Biology, Keio University School of Medicine, Tokyo.

Japan's Human Genome Programme consists of five topics: human genome mapping, human genome function, DNA technology, bioinformatics, and nonhuman genomes. Each has five to fifteen researchers engaged upon it and is directed by co-leaders. Over 40 research laboratories from various universities and institutes are engaged in the programme.

1. The human genome mapping group performs linkage mapping and physical mapping of human chromosomes, and develops improved mapping techniques for rapid isolation and characterization of disease-related genes. This group is also interested in the analysis of such chromosome domains as telomeres and centromeres. Construction of linking clones, YAC libraries, and P1 libraries for selected chromosomes is under way.

2. The human genome function group deals with construction of tissue — and/or chromosome-specific cDNA libraries of high quality for the analysis of genome sequences encoding proteins.

3. The DNA technology group deals with separation of giant DNA segments, cloning of large DNA segments, specific cleavage of DNA and new approaches to speed up DNA sequencing.

4. The bioinformatics group deals with maintenance of the existing DNA data-base (DNA data-base of Japan), establishment of new mapping and genome data-bases, development of a new "knowledge data-base", various kinds of software for data analysis, and creation of a relatives data-base.

5. The nonhuman-genomes group deals with the genome analysis of *E. coli, Saccharomyces cerevisiae, C. elegans, Drosophila melanogaster, and M. musculus.* The *E. coli* genome sequencing project is separately supported by the Ministry of Education, Science and Culture.

The planning committee of the Human Genome Programme is well aware of the social and ethical problems and is considering the establishment of a special group to be responsible for this aspect.

Several other genome-related programmes also receive governmental support. These include: the Riken project, the Yoken project, the Genosphere project, and the Human Frontier Science Programme.

1. The Riken project stems from the "Wada project (1981-1988)", which was re-structured in a new programme in 1989 owing to a special report by a Council for Aeronautics, Electronics and other Advanced Technologies (Koudenshin). The Riken project receives 200 million yen ($1.3 million) a year from the Science and Technology Agency for the complete sequencing of human chromosome 21. This project includes several research groups from universities and private companies.

2. The Yoken Genome Project has just begun as a part of the "Silver Science Programme", which is supported by Kosei-sho (the Ministry of Health and Welfare). The initial budget is 300 million yen ($2 million) a year. The project focuses on the analysis of human disease genes from a clinical point of view, using currently available strategies and technologies. Yoken plans to enforce the existing cell and gene bank (Japanese Cancer Research Resources Bank) through this programme.

3. The Genosphere Project is supported by the Research Development Cooperation of Japan under one of its programmes called Exploratory Research for Advanced Technology. "Genosphere" is analogous to "Biosphere" and implies the genome's interactions and surroundings. This project receives 2.5 billion yen ($16.6 million) for five years and recruits 20-30 researchers throughout the world to study the following research topics: analysis of chromosome structure by means of a laser microdissector; analysis of chromosome function by microinjection of the dissected chromosome fragments; and analysis of the dynamic behaviour of chromosomes during the cell cycle by means of new graphic technologies.

4. The Human Frontier Science Programme has been established through the Science and Technology Agency and the Ministry of International Trade and Industry to promote international cooperation in basic scientific research. The participating countries are the Economic Summit countries: Canada, France, Germany, Italy, UK, USA and Japan. Its headquarters was founded in Strasbourg in 1989 and an International Peer Review Committee was formed to review applications for research grants, fellowships and international workshops. The 1989 budget was 1.9 billion yen ($13.6 million) from the Agency and 1.4 billion yen ($10 million) from the Ministry. The Programme emphasizes support in special research areas, including brain and biological functions. This does not exclude human-genome-related research: the Programme sponsored an International Workshop on Molecular Approaches to the Human Genome held in Oiso, Japan in March 1989, attended by over 100 scientists, including 40 foreign scientists. The meeting was considered the first successful scientific meeting on the human genome.

The Ministry of Agriculture, Forestry and Fisheries has created a Rice Genome Project, and very recently announced a joint project between the US and Japan for the use of fifth generation supercomputers in genome data analysis. The various human genome projects have been funded by separate governmental organizations and official coordinating efforts have been very limited. Working together on the human genome programme is critical for its success.

Applications of the information generated by the genome project are undoubtedly enormous, and cost-sharing among the nations has been a great concern. Japan's contribution to the human genome project is relatively small, but there is increasing enthusiasm in the nation and larger financial support is expected. Our immediate task is to level up and expand the genome field as quickly as possible. We, like others, see that the human genome belongs to the world's people, not to a single nation, and therefore international cooperation is critical. The Human Genome Organization (HUGO) has been established and is expected to provide timely advice and facilitate international cooperation. To aid this, HUGO Japan (Professor Matsubara, Vice President of HUGO, and 11 other HUGO members) has been making efforts at fund-raising.

35

II. Physical mapping and fine structure analysis of the human genome by the use of flow-sorted human chromosomes: the Keio strategy

We are interested in physical mapping of the smallest human chromosomes, 21 and 22, which contain DNA of approximately 50 Mbp.[1] These chromosomes contain a number of clinically important disease genes. For example, chromosome 21 is associated with Down's syndrome and Alzheimer's disease, and chromosome 22 with neurofibromatosis and Ewing's sarcoma.[2] Identification and isolation of these genes can be facilitated by the information and materials generated by physical mapping and genetic linkage mapping. To achieve this goal we have recently devised a comprehensive "top down" strategy for physical mapping and fine structure analysis of a particular chromosome by the use of flow-sorted chromosomes.

In our "Keio strategy" for physical mapping of human chromosomes 21 and 22, we isolate intact chromosome 21 (or 22) by flow sorting and utilize the isolated chromosomes for the following three different purposes: 1. To analyze size distribution of DNA segments after digestion with the restriction enzyme *Not*I on pulsed field gel electropheresis (PFGE). The DNA segments will be isolated and used to construct a clone library specific to a particular chromosomal region by shotgun PCR. 2. To clone *Not*I DNA fragments into yeast artificial chromosome (YAC) vector. Also, the flow-sorted human chromosomes are used to construct a clone library by means of P1 phage vector. 3. To construct a chromosome 21 (or 22)-specific linking library. The linking library will be used as a probe to identify two adjacent *Not*I DNA segments on PFGE as well as DNA segments contained in the YAC library. YAC clones will be analyzed by constructing a more detailed restriction map. The YAC clones will also be used for "giant walking" and fluorescent *in situ* hybridization.

For chromosome sorting, metaphase chromosomes were prepared by the polyamine-digitonin method from lymphoblastoid line GM130B, which has an apparently normal male karyotype.[3] These chromosomes were stained with propidium iodide and sorted *via* a FACS440 sorter equipped with a single argon laser. Sorting speed is usually about 80 chromosomes 21 (or 22) per sec. Chromosomes 21 and 22 can be sorted separately as a single peak. Approximately 500,000 chromosomes were collected, embedded in a low melting-temperature agarose gel block, and treated successively with proteinase K/sarkosyl and the restriction enzyme *Not*I. The digested DNA fragments were analyzed by PFGE followed by Southern blot hybridization with the human *Alu* repetitive sequence as a probe. The analysis revealed over 30 distinct large DNA segments, ranging from 50 kb to longer than 2.5 Mb in size. Non-specific fragmentation by shearing or digestion with endogenous nuclease was minimized, demonstrating the intactness of flow-sorted chromosomes. Thus, we established sorting conditions to obtain intact human chromosomes 21 and 22 suitable for physical mapping. This was made possible by a number of technical improvements.[4,5]

In a separate experiment, we analyzed a centromeric region by using chromosome 21 (and 22)-specific alphoid DNA probes. There are at least two giant DNA segments of 4-5 Mb for each of these two chromosomes. Thus, the centromeric *Not*I segments containing alphoid sequence of chromosome 21 (and 22) may consist of more than 10Mb.

Next, we have attempted to clone *Not*I-DNA fragments from flow-sorted human chromosome 21 into YAC vectors. The original pYAC55 vector provided by M. Olson[6] was modified in such a way as to rapidly subclone both ends of the human DNA inserts. Approximately 10 million chromosomes 21 (or 22) were used for YAC cloning. These chromosomes were embedded in a low-melting temperature agarose block and treated with proteinase K and the restriction enzyme *Not*I. The agarose block was melted at 68°C and the DNA fragments were ligated to pYAC55S2, according to the method of Burke *et al*. We have isolated about 200 Ura[+] Trp[+] red transformants. This first series of YACs contained predominantly human ribosomal RNA genes of relatively small sizes ranging from several kb to 40 kb. This was accounted for by the presence of about 80 copies of rRNA genes on each of the human chromosomes 21 and 22.[7] Improvement of YAC cloning is under way.

Once *Not*I DNA fragments are identified on the pulsed field gels and cloned into YAC vectors, they must be aligned in order. Ordering of *Not*I DNA segments can be accomplished by a chromosome 21 (or 22)-specific *Not*I linking library.[8] Each linking clone provides two RNA probes for each of two sides of the *Not*I site. We are in the process of utilizing these linking clones to order the *Not*I DNA fragments that are detected either by PFGE or in the YAC library.

An interesting new observation is related to the ribosomal RNA genes. In the human genome there are about 800 copies of rRNA genes which are organized as tandem repeat units and distributed in clusters over the short arms of all five acrocentric chromosome pairs. The rRNA genes are located in the secondary constriction of these chromosomes and active transcription is detected as silver-stained nuclear organizer regions. Our studies using flow-sorted acrocentric chromosomes provided molecular evidence that there are differences in the number of rRNA gene copies on different chromosomes and in different individuals, and further suggested that differences in the structural organization of the rRNA gene clusters on different chromosomes are reflected by the degree of methylation.

There is an important consideration with regard to the flow-sorted chromosomes, namely that these sorted chromosomes are pure but mixed with maternal and paternal homologs, and may not be ideal for physical mapping and future DNA sequencing, owing to individual DNA polymorphisms. To solve this potential problem, we have improved the sorting method to isolate only single homologs of chromosome 21 (or 22), by the use of lymphoblastoid cells carrying unique translocation chromosomes. Comparison of the DNA restriction patterns of normal and translocation chromosomes provided direct information on the particular regions of translocation.

Although our work is only at its beginning, the Keio strategy has proven to be feasible. The chromosomes flow-sorted under our sorting conditions are intact and therefore suitable for the identification of giant DNA segments corresponding to specific chromosomal regions. We are developing a method to construct a clone library specific to a particular human chromosomal region by means of shotgun PCR. Various clone libraries to be generated should eventually allow for physical mapping of these chromosomes and functional analysis of the selected chromosomal regions. Isolation of potential disease-related genes will be attempted in reference to a genetic linkage map.

Acknowledgement

This work was supported in part by a grant-in-aid for the Human Genome Programme from the Ministry of Education, Science and Culture and the Science and Technology Agency.

References

1. Southern, E.M. Application of DNA analysis to mapping the human genome. *Cytogenet. Cell Genet.* 1982; 32:52-57.
2. McKusick, V.A. *Mendelian Inheritance in Man.* Johns Hopkins University Press, Baltimore, 1989.
3. Sakai, K. et al. Isolation of cDNAs encoding a substrate for protein kinase C: Nucleotide sequence and chromosomal mapping of the gene for a human 80K protein. *Genomics* 1989; 5:309-15.
4. Minoshima, S. et al. Isolation of giant DNA fragments from flow-sorted human chromosomes. *Cytometry* 1990; 11:539-46.
5. Shimizu, N. and Minoshima, S. Gene mapping and fine structure analysis of the human genome using flow-sorted chromosomes. In: *Advanced Techniques in Chromosome Research.* (ed) K.W. Adolph, Mercel Dekker Inc. In press.
6. Burke, D.T. Cloning of large segments of exogenous DNA into yeast by means of artificial chromosome vectors. *Science* 1987; 236:806-12.
7. Shimizu, N. et al. *Variation in the organization of the rRNA gene clusters among flow-sorted human acrocentric chromosomes*, Cold Spring Harbor Conference of Genome Mapping and Sequencing, New York, Abst. P. 2 (May 1990).
8. Shimizu, N. Molecular strategies for physical mapping and fine structure analysis of flow-sorted human chromosome 21. *Proceedings of the First International Conference on Electrophoresis, Supercomputing and the Human Genome.* (ed) C. Cantor and H. Lim, World Scientific. In press.

THE SOCIAL GEOGRAPHY
OF HUMAN GENOME MAPPING

Bartha Maria Knoppers* and **Claude M. Laberge****

The quest of the mapping of the human genome cartography is one that elicits the expression of the most extreme public phobias and polemics surrounding scientific advances. Yet, unlike the new reproductive technologies or the discovery of the human immunodeficiency virus (HIV), which confronted the public with a scientific *fait accompli*, the scope and timing of the project is such that the more complex ethical, legal, social and cultural issues[1] can be addressed before there is widespread application of the knowledge gained. Nonetheless, the fall-out of delayed public consultation and reaction to both assisted conception and AIDS has aggravated public fears and scepticism with respect to the new science of genome mapping.[2]

Indeed, the rationale and usefulness of mapping the human genome can be questioned in the light of the fact that the exact DNA alteration of sickle-cell anemia, to take an example, has been known for at least 30 years but its associated serious health problems have not been solved.[3] Also, there is some concern lest important genetic and other research be neglected because of the cost of the human-genome project. This critique of the project, when coupled with the concerns about its possible societal or eugenic applications, has led to a much more cautious approach, particularly in Western Europe.[4]

At the same time, as with all scientific advances, the old catch-phrases "playing God", "interfering with nature", and "going down the slippery slope" have resurfaced, along with two new fallacies specific to the social geography of this area: the first, that science is ethically neutral, and the second, that genetics is the answer — the ultimate, universal, technological "quick-fix".[5] To these should be added the "capacity fallacy" and the "necessity fallacy", underlying the phobia of technological imperialism.[6] Scientific publishing in the area of molecular biology has added another "eternal truth", the "assertive sentence titles (AST)", since titles now "assert conclusions rather than causes".[7] Finally, like the initial fear accompanying the introduction of roentgenography, revealing our skeletal structure on an X-ray[8], the genetic revelation of the secrecy of biological relationships (cf. non-paternity) and of personal health status is seen as threatening to the individual.

The discussion in the literature on this matter uses the language of polemics, reflecting the poles of perception separating different ideologies. These poles are expressed in the debates on evolution versus creation and

---

*   Associate Professor, Faculté de droit, Centre de recherche en droit public, Université de Montréal, Canada.

** Professor of Genetic Medicine and of Pediatrics, Laval University Medical Centre and Quebec Network of Genetic Medicine, Québec, Canada.

nature versus nurture, genetic determinism[9] versus genetic chance, and more recently, genetic uniqueness[10] versus genetic immutability.[11] Fundamentally they represent the extreme monocular, dualistic views of human nature.

Since we have the time to reflect and to inform and consult the public, should we not employ stereoscopic vision? One of the two pictures of the social geography of the human-genome mapping project, taken from differing angular views, could be superimposed on the other and the two seen as one picture. Such a stereoscopic approach, however, requires more than the discarding of old catch-phrases and fallacies, or the avoidance of dualisms. It requires education of the public in the language of risks and probabilities, of probes and prediction.[12] It requires the realization that the new genetics will not be the science of manipulation, elimination or imperialism but rather the art of revelation and communication of genetic information.[13]

Taking into consideration, then, these phobias and polemics surrounding the social geography of the human-genome mapping project, this paper will adopt an isometrical perspective, beginning with the gross physical structure — that is, the topography — of the map, proceeding to the genome information of its demography, followed by a study of the translation of this information into genetic epidemiology, and, ultimately, the application of such knowledge to the citizenry so as to form the basis of a new social contract[14] (Fig. 1).

Genetic topography

Genetic topography (Fig. 2) demonstrates the anatomical genetic structure of a species. The genic cartography of the human species will chart on sets of maps the physical location of genes and DNA sequences, of diseases, of traits and of multi-linkage groups (genetic maps of male and female recombinant fractions). Genic cartography details the genome structure, allocates sequences on chromosome sites and marks polymorphisms at these sites. It shows that man as a species is basically animal, DNA being the ''stuff'' of life.[15] Human genes have homologies with genes of other species, or at least evolutionary ties. Therefore, some of the human-genome mapping will be aided by animal studies and by using animal sequences as probes for the identification and isolation of genes. The study of gene expression in transgenic animals will serve as one of the bases for gene therapy.[16] Finally, even the non-coding sequences of DNA serve in the cartography as walking/jumping sites.[17] They also contain regulatory sequences. In short, as the genome unfolds it will provide the scientific proof of evolution. It will put man in the ''stream'' of living matter. How then can we read this level of the human-genome map?

Firstly, the human-genome map will be anonymous and will demonstrate membership in the human species irrespective of race or colour.[18] Secondly, as human genetics knows no political or social boundaries, the information gained will apply to everyone, irrespective of individual differences.

Figure 1.

THE SOCIAL GEOGRAPHY OF HUMAN GENOME MAPPING

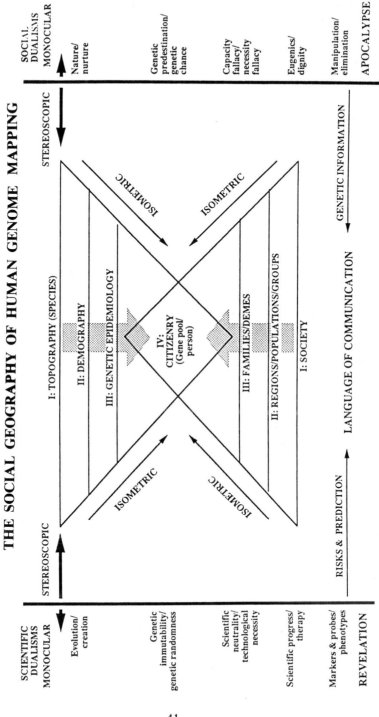

41

Indeed, this information will have to be characterized to populations and to individuals.

Thirdly, at this level of cartography, the collaboration must be international and raises the issue of project control and the role of such international organizations as the Human Genome Organization (HUGO).[19] Will the traditional scientific sharing of technology[20], of cell lines, of DNA from reference families, of probes and of data continue, or will control take the form of applications for patents in view of the commercial implications inherent in the biotechnology spin-offs?[21]

Fourthly, at the topographic level of mapping might it not be better to consider the information gleaned as the common heritage of mankind rather than as a commercial commodity?[22] The basic common factors are that: use of the information must be for purposes consonant with peace; access must be open to those who have a right to it, while the rights of others must be respected (therefore, responsibility for abuse); sharing must be equal; and, owing to its indivisible character, the genetic heritage must be administered in the interests of all for the common good. This international concept stems from the need to prevent ownership of things of communal interest and to preserve things that are of international interest for future generations. Like the legal concept of a trust under private law [once limited to "things" *(res)* and now extended to interests and rights], the concept is akin to that of a "public" trust which assists in the transmission of property or interests from one generation to the other.[23] In essence, the question involves a determination of "the extent to which our collective gene pool is public property, which we hold in trust for the future, and the extent to which the very personalized packages into which it is subdivided preclude treating it as a public resource".[24]

In 1987, the United States Office of Technology Assessment alluded to the possibility of Congressional action which would recognize "that any cell line be presumed to be in the public domain", thus barring anyone from claiming property rights to these products.[25] Similarly, the U.S. National Research Council Committee on Mapping and Sequencing the Human Genome recommended in 1988, "that human genome sequences should be a public trust and therefore should not be subject to copyright".[26] This possibility was also raised by the American Society of Human Genetics Committee on Mapping and Sequencing when it solicited comments from its members on whether a future statement should include the public-trust position — that is, "an expression of belief that the human genome sequences would be a public trust and therefore not subject to copyright".[27] Such an approach need not be seen as contradictory to the recent statement of the Ad Hoc Committee on DNA Technology of that Society with respect to individual deposits of DNA in DNA repositories. This position maintained that "banked DNA is the property of the depositor unless otherwise stated".[28]

It could be argued that one's individual control over the uses of one's tissues and cells is not incompatible with the notion of the human genome

Figure 2.

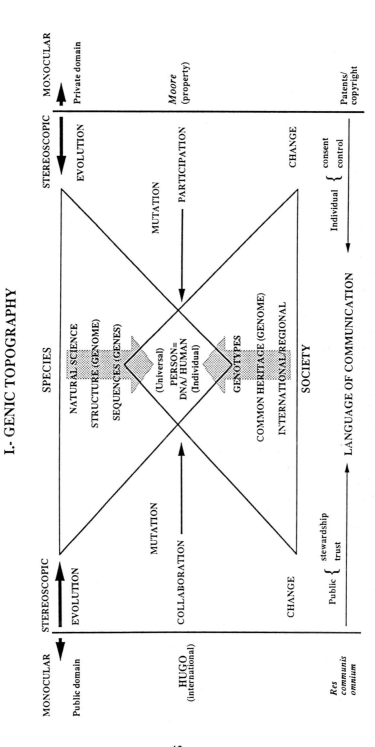

I.- GENIC TOPOGRAPHY

itself being considered a public trust. It should be possible to balance the notions of individual "genetic control" with public stewardship of the larger "gene pool" or human genome without recourse to property law.[29] Such an approach would parallel the existent biological reunion of the individual pattern and gene-pool source of the human person. It would provide for individual control and yet encourage the individual to participate in the familial context.

Finally, there have been both regional[30] and international appeals[31] for legal protection of the human genetic heritage on the basis of respect for the inherent dignity of the human person.[32]

Genetic demography

Turning from species to populations, the information obtained from mapping the human genome will be applied to genetic demography (Fig. 3). Demography is the study of population growth through life cycles and through migrations. It is a human science like history, geography and philosophy.

Genomic information will give the necessary content for a new form of population genetics, that of genotypic rather than phenotypic markers.[33] This population genetics will use gene markers and restriction fragment length polymorphisms (RFLPs) as polymorphic tools. It will provide the knowledge necessary to build reference populations for the validation of DNA testing.

Furthermore, the quantity and diversity of neutral mutation[34] in coding and especially in non-coding areas will provide enough polymorphisms for population studies and eventually will demonstrate the uniqueness of individuals in populations. Genomic information emanating from sequences will also contribute to knowledge about embryology and development, on etiology of diseases, on aging processes and perhaps on behaviour in the larger sense of neuro-science. Population information on sequences will demonstrate that mutations are not black and white phenomena. There is a variability of sequences at genic sites just as there is variability in phenotypic expression of diseases, e.g. Tay-Sachs[35] (gene mutations) and myotonic dystrophy[36] (variability of expression). Finally, new information about and from the sequences will furnish new tools for research into, for example, homologies between receptor superfamily and oncogenes,[37] and into the possibilities of imprinting.[38] In short, genetic demography is the basis for genetic medicine, the basis of genetic physiology studies of the systems, processes and uses of sequence information in "reverse" genetic medicine.[39] How then shall we read this level of the map?

Genetic demography gives information on the sites and brings with it a shift in the paradigms of research and medicine towards genetic medicine.[40] Moreover, as with genetic topography, the possible impact of these changing paradigms on transfers between banks or repositories of cells, familial data, DNA, tissues and information, and on the withholding of information from publication, cannot be overemphasized.

Figure 3.

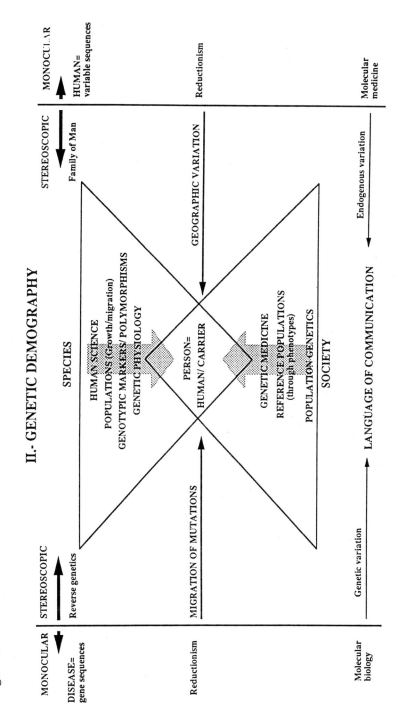

II.- GENETIC DEMOGRAPHY

45

Above all, there is the danger of genetic reductionism to sequences — that is, that humans are seen as variable sequences and that diseases are caused only by gene sequences. Adherence to such a reductionism would reduce medicine to molecular biology, where there is a tendency to equate sequence with disease. Such an attitude would constitute a misreading of the genetic demography map and ignore genetic diversity and complexity, even on a population level.

Genetic epidemiology

Genetic epidemiology seeks to discover the incidence of genetic problems in populations or groups and to gather samples for the study of the expression of genetic deseases or traits[41] for purposes of clinical research. Genetic epidemiology will provide the tools of measurement for the translation of genome information to target populations, to ethnic groups and to families. In this sense, it is a social science, like economics and political science. This translation of information about the sequences onto the political and cultural structures of populations and groups will serve as the basis of preventive and predictive medicine. It will use the information as tools for screening, separation, classification, measurement of incidence and prevalence, clustering and targeting, for the planning of genetic health-care resources, and for the study of the feasibility of providing genetic services to populations (Fig. 4).

Classical epidemiology seeks to identify causes of disease in the environment within the well-known approach of "abnormalcy" being due to lifestyles or geography. Knowledge of the sequences however will allow genetic epidemiology to look for the variation and the expression of an endogenous cause (gene) in the environment. How then can we use this mapping technique?

On the one hand, it is possible that instead of all the population being considered at risk, the act of naming the mutations in the population could eventually make the notion of "abnormalcy" disappear. On the other hand, until research can name a great number of mutations, the widespread ignorance of genetics may result in an increase of social discrimination.[42] It is also a potential source for the stigmatization and ostracism of populations. There is the danger that target populations will be perceived as diseased and that "at risk" populations will be correlated with "defective" populations.[43]

What is the State interest in such information for the planning of screening programmes? The traditional cost-benefit analysis on which the universal screening of newborns or pregnant women could be justified fails in the genetic context.[44] The causes no longer being solely in the environment but in the genealogy, risk identification crosses the traditional division of medicine and epidemiology. The result may well be a new economic rationale replacing the cost-benefit basis. For instance, the rarity of certain genetic "orphan" diseases could lead to elimination by selection rather than by spending money on research and treatment. Similarly, prenatal

Figure 4.

III.- GENETIC EPIDEMIOLOGY

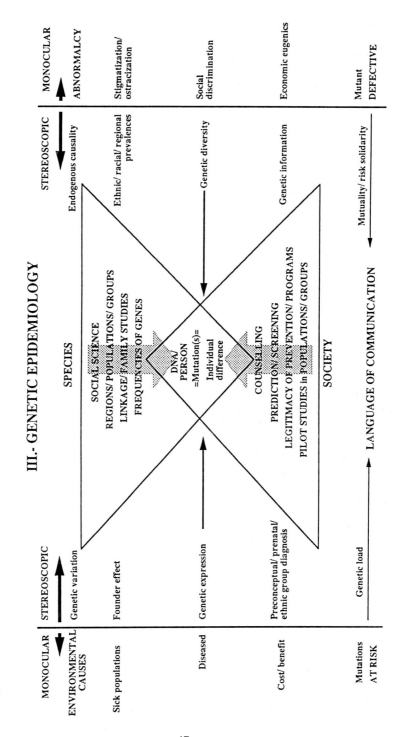

47

or pre-conception diagnosis could be encouraged so as to avoid long-term costs to the State.[45] This rationale, disguised as responsible planning and priorities of resource allocation, could lead to the most dangerous and subtle form of eugenics, economic eugenics. In the absence of widespread genetic information and education for the public, this could well come about.

Choices have to be made, however. Populations, regions, ethnic groups or families should be consulted as to the legitimacy of research projects after the completion and scientific documentation of pilot studies — that is, before clinical research or prevention programmes are even offered.[46] Such genetic epidemiology may well demonstrate the fellowship of diseased individuals in populations, since there is a genetic load to any development in the demography of a population. Consequently, there is an ensuing social responsibility to be shared by individuals, a form of risk-solidarity or mutuality inherent in genetic decision-making at this level.[47] Are these ethical values of kinship and risk-solidarity or mutuality applicable to populations, targets or clusters, large demes or families and regions?

Ultimately, choices should still be based on individual informed decision-making after genetic counselling as to risk status. Any eugenics then would be the sum total of individual decision-making following counselling. These individuals, after public consultation and consent as to the legitimacy of genetic preventive programmes, would have thus participated in the social choices inherent in genetic epidemiology. This brings us to the final dimension of social geography, the citizen, the individual living in a given society.

The citizenry

We are now at the level of the individual patrimony, for genetic heritage is both collective and individual, in the sense that the personal genotype is the totality of inherited characteristics drawn from the gene pool at a given time.[48] This "genecity"[49] of the human person is situated in the realm of individual rights and freedoms. It will name the individual genetically but, contrary to common perception, the nomenclature will itself be in constant mutation (Fig. 5).

The application of sequence information, as translated by epidemiology and now by genetic medicine, to individuals presumes access by the individuals to such information and services. Even though at present risk assessment is mainly performed by linkage analysis, molecular geneticists must prepare themselves for the transmission of knowledge to individuals and families. Families are needed to find and validate predictive tests and to answer questions about disease or carrier status. It is ironic that, at a time when genealogical family has been replaced by the social concept of the family, there is a reconstruction of the biological family for genetic purposes.

An even greater transformation is that, as we have shown, genome mapping will allow us to demonstrate not only genetic diversity and individuality within a given biological family but also kinship, the interrelatedness

Figure 5.

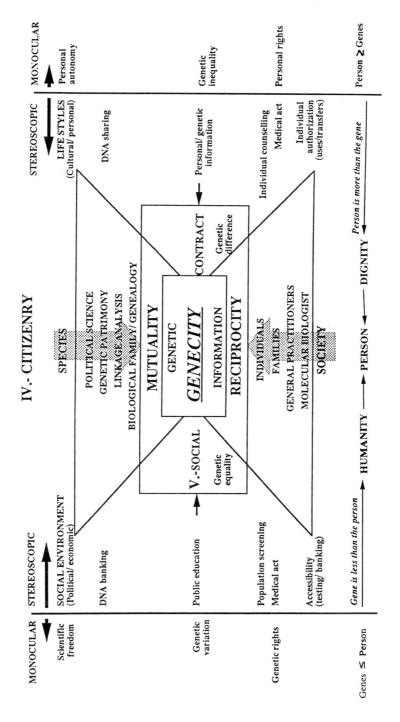

IV.- CITIZENRY

MONOCULAR

STEREOSCOPIC LIFE STYLES (Cultural/ personal)

MONOCULAR

Personal autonomy

SPECIES

DNA sharing

Genetic inequality

Personal rights

POLITICAL SCIENCE
GENETIC PATRIMONY
LINKAGE ANALYSIS

Personal/genetic information

BIOLOGICAL FAMILY/ GENEALOGY

MUTUALITY

CONTRACT

Individual counselling

Medical act

GENETIC

Genetic difference

Individual authorization (uses/transfers)

GENECITY

INFORMATION

INDIVIDUALS

Person ≥ Genes

FAMILIES

RECIPROCITY

GENERAL PRACTITIONERS

V.-SOCIAL

MOLECULAR BIOLOGIST

SOCIETY

Genetic equality

SOCIAL ENVIRONMENT
(Political/ economic)

STEREOSCOPIC

DNA banking

Public education

Population screening

Medical act

Accessibility
(testing/ banking)

MONOCULAR

Scientific freedom

Genetic variation

Genetic rights

Person is more than the gene

HUMANITY ⟶ PERSON ⟶ DIGNITY

Gene is less than the person

Genes ≤ Person

49

of man. How then can each citizen as a member of a family, of a given society, use this highly personal and yet common map?

Firstly, molecular biologists should not be allowed to monopolize genetic information. Knowledge of human genetics should be part of the armamentarium of the medical practitioner. Both physicians and patients are at risk if a commercial, property approach is encouraged; this would constitute a transition in science and medicine to economic interests. Indeed, it is the relationship between the ordinary physician and the patient that is at the greatest risk from the genome map. The contractual physician-patient relationship could be synergized by the communication of this genetic information whithin its framework and prerogatives. Also, self-determination, autonomy, procreative choice and lifestyle choices belong here.[50] Thus, testing and communication of genetic information must be recognized as a medical act, not a molecular-genetic act.[51]

Secondly, at this level of the map it is imperative that genetics be exempt from claims of rights in relation to the individual genetic heritage.[52] To be otherwise would involve the risk of genetic recriminations and claims within families and between generations on the basis of quality-of-life choices, and this all the more so in the current social construct. For that same reason, the physician must be exempt from legal obligations to third parties. The communication of genetic information may in some instances constitute a moral obligation but the communication of genetic information or even the search for other family members in the construction of the pedigree remains the responsibility of the individual within the familial context.[53] In this familial context, "the genetic nature of the data may give rise to modifications in the usual rules".[54]

Thirdly, the patients must be in a position to control and authorize any use, transfer, or communication of genetic material and genetic information to themselves or to third parties.[55] Failure to ensure this control through the presentation of options for banking, for research, for the sharing of data or tissues or for the communication of results[56] would encourage the predominance of economic-property concepts, of patent law[57] or of statutory definitions of the human being.[58]

Finally, without the genetic education of the public we cannot expect or exact the participation of the citizen in the communication of genetic information, in the social planning of prevention programmes, in the mass screening of populations for treatable disorders or, one day, in programmes providing at-risk information with regard to susceptibility and reproduction choices.[59] Only an informed citizenry will and can participate, and then only on the basis of a new social contract.

Conclusion: The new social contract of genomics

The mapping and understanding of the human genome could change the former basis of Rousseau's social contract.[60] According to Rousseau, the contract was necessary to compensate for basic inequality. However, if genetic difference and diversity were to be understood as the basis for equality, genetic knowledge would lift what Rawls has called the "veil of ignorance".[61] For Rawls, the veil of ignorance hides the given natural distribution from those who choose the principles of justice. Now individual knowledge of genetic difference could foster equal rights and equal obligations, since everyone would be seen as genetically different and thus equal in diversity. Such an approach to the application of knowledge gleaned from the understanding of the human genome could herald the end of medical and social discrimination. The address of the sequence becomes secondary. Health then becomes adaptation within one's own genetic difference and variation with others, and within a new social contract based on the appreciation of difference. This new social contract could be founded on the principle of reciprocity of exchange in the physician-patient relationship and on the principle of mutuality in participation within the familial context. The human genome map creates a model of the human species but does not make the map an individual structural model (Fig. 5).

The social geography of the human genome is both collective and individual. Recognizing the universal basis of our common genetic heritage will serve to ensure the appreciation of its international nature and the avoidance of individualistic property concepts. The social geography of the human-genome mapping project requires a stereoscopic vision of its topography so as to avoid the classical dualisms. The transmission of demographic knowledge of sequences to populations through genetic epidemiology must avoid the equation of the sequence to the disease.

Finally, the education and participation of medical practitioners and of the public are crucial. Such education and participation will serve to situate the locus of the communication and control of individual choices in genetic medicine, not in the molecular biology laboratory but in the physician-patient relationship. This relationship is the ultimate insurance against state eugenics and against the emergence of the language of "genetic rights". Only then will the equation between genetic revelation and apocalypse, between genetic information and discrimination, between the person and the disease, be erased.

The person is not the gene or the disease. The ignorance of genetic variability, complexity and mutation that translates into the social constructs of "abnormalcy" or of "defective" must be transformed into "equality through genetic difference". Only then will the stereoscopic vision of the human-genome map allow for the equation of humanity with citizenry.

References

1. See generally: International Conference on Bioethics (Rome April 10-15 1988) *Human Genome Sequencing: Ethical Issues.* Brescia, Italy: Clas International, 1989.
2. See, for example: Council for Responsible Genetics: Position paper on human genome initiative. Boston, MA, 1989.
3. Weatherall, DJ. Mice, man and sickle cells. *Nature* 1990; 343:121; Luzatto L, Goodfellow P. News and Views: (Sickle-cell anemia) A simple disease with no cure. *Nature* 1989; 337:17.
4. Knoppers BM. Genetic heritage: The international debate. In: BM Knoppers & CM Laberge (eds) *Genetic Screening: From Newborns to DNA Typing.* Excerpta Medica, Int'l Congress Series 901. Amsterdam: Elsevier, 1990:257.
5. Boone CK. Bad axioms in genetic engineering. *Hastings Center Rep.* 1988; 18:9. Taking these in turn, the author posits in reply to the first that we must make intelligent choices, accepting or rejecting genetic options. With regard to "nature", these choices should be based on some human conception of what is natural, not on a naturalistic definition of what is human. Thirdly, the fear of "slipping" as a reason to halt genetic engineering is equivalent to denying afflicted individuals therapy "on the ground that we cannot make distinctions between remedial germline alterations and eugenic enhancements. [It] indicates a lack of trust in the human ability to act discriminately on the basis of ethical classifications". The author maintains that the ethical neutrality of science is a myth since "the scientist *qua* scientist does not really exist". For him, the scientist always remains a moral agent. Finally, the claim that we have achieved or will achieve genetic omnipotence is an underestimation of genetic complexity and an example of human hubris. According to the author, "long before teleological thirst deteriorates into technological lust, it will need tempering by the acknowledgement of human finitude and by the willful determination to resolve problems by means that realize human integrity, not ones that undermine it".
6. Fletcher J. *The Ethics of Genetic Control: Ending Reproductive Roulette.* New York: Doubleday, 1974 at 5ff. on the errors of the capacity fallacy (we can do something, so we ought to) and the necessity fallacy (we can do something, so we will).
7. Rosner J. Reflections of science as a product. *Nature* 1990; 345:108. "The AST has the form of an eternal truth". (The recent news release of the discovery of "the" gene responsible for alcoholism is a good example of this trend).
8. Scriver CR and Holtzman NA. Future types of screening: A prospectus. IV.-Ethical and legal safeguards. In: BM Knoppers & CM Laberge (eds), *supra,* note 4, 330.
9. Amsterdamski S, Atlan H, Danchin A, Ekeland I, Largeault J, Morin E, Petitot J et al. *La querelle du déterminisme (Philosophie de la science d'aujourd'hui).* Paris: Gallimard, 1990.
10. Jacquard A. *L'Héritage de la liberté: De l'animalité à l'humanitude.* Paris: Seuil, 1986. See also Rose S, Kamin LJ, Lewontin RC. *Not in Our Genes: Biology, Ideology and Human Nature.* Suffolk: Chaucer Press, 1984:51.
11. Wilson EO. *Sociobiology: The New Synthesis.* Cambridge, Mass.: Harvard University Press, 1975.
12. Holtzman N. Recombinant DNA technology, genetic tests and public policy. *Am J Hum Genet* 1988; 42:624.
13. See generally, Knoppers BM. *Genetic Heritage and Human Dignity.* Study Paper, Law Reform Commission of Canada, 1990.
14. Knoppers BM, Human genetics, predisposition and the new social contract. In: International Conference on Bioethics (Rome April 10-15, 1988) *Human Genome Sequencing: Ethical Issues: Diagnosis and Prediction: Ethical Issues related to Society.* Brescia, Italy: Clas International. 1989:168.
15. Dagognet F. *La Maîtrise du vivant.* (Histoire et Philosophie des Sciences). Paris: Hachette, 1988.
16. Animal models are not always applicable or may be problematic, see *supra* Weatherall, note 3. See also Erickson RP. Minireview: Creating animal models of genetic disease. *Am J Human Genet* 1988; 43:582.
17. Goodfellow PN. (Cystic fibrosis) Steady steps lead to gene. News and Views, *Nature* 1989; 341:102.

18. The genetic pattern of each individual is unique, and yet there cannot be genes in the present population that did not exist at the beginning of the species of "modern man", about 30,000 years ago (barring mutation). This universal, ancestral filiation makes the human gene both individual in expression and supranational in origin. In other words, while heterogeneous in appearance, the human race is basically genetically homogeneous, and any differences are due to recent (30,000 years) and local adaptations rather than to historical diversity. This is due to the fact that the evolution of our collective genetic patrimony requires hundreds of generations to effect a significant change. See generally, Stringer CB, Andrews P. Genetic and fossil evidence for the origin of modern humans. *Science* 1988; 239:1263. Also *Langaney A. La diversité génétique humaine: considérable et mal connue.* In: *Génétique, procréation et droit.* Paris: Actes Sud, 1985:349.

19. Anderson GC. Howard Hughes gets HUGO off the ground. *Nature* 1990; 345:100. Regarding principles concerning collaboration and participation in HUGO, see Cantor CR. HUGO physical mapping. Correspondence: *Nature* 1990; 345:106.

20. Swinbanks D. Human genome project a cause of friction. *Nature* 1989; 342:463; Coles P. News: Human genome: Keeping them guessing. *Nature* 1990; 343:579; Ewing T, Swinbanks D. News: Human Frontier Science Program: Turned back at the door. *Nature* 1990; 345:9.

21. *Moore v. The Regents of the University of California*, 202 Cal. App. 3d 1230, 249 Cal. Rptr. 494 (1988): Sup. Ct. Calif. File no: S006987 (July 9, 1990).

22. The remaining discussion under II.-Genetic topography to note 29 is taken from Knoppers, *supra,* note 13.

23. De Jager K. Claims to cultural property under international law *L.J.I.L.* 1988; 1:183.

24. Neel J. as cited In: LB Andrew. *Medical Genetics: A Legal Frontier.* Chicago: American Bar Foundation, 1987, at 250.

25. Office of Technology Assessment, *New Developments in Biotechnology: Ownership of Human Tissues and Cells.* Congress of the United States, 1987; Hermitte MA. *Histoires juridiques extravagantes: la reproduction végétale,* at 40 and also by the same author: *Le concept de diversité biologique et la création d'un statut de la nature,* at 238, as well as C. Labrusse-Riou C. *Servitude, servitudes.* Texts found In: *L'homme, la nature et le droit.* Paris: C. Bourgois, 1988, at 308.

26. *Ibid.,* at 91.

27. Short E. Proposed ASHG position on mapping/sequencing the human genome. *Am. J. Hum. Genet.* 1988; 43:101, at 102.

28. DNA banking and DNA analysis: points to consider. (1988) 42 (5) *Am. J. Hum. Genet.* 1988; 42:781, at 792.

29. "Human Genome Analysis" programme tabled for the European Parliament, COM (89) 532, Nov. 13, 1989, at Section 3.2. With respect to all contracting parties under this European programme:
   "there shall be no right to exploit on an exclusive basis any property rights in respect of human DNA".

30. Dufour A, Hondius FW, *Vers une convention bio-médicale. Forum Conseil de l'Europe.* Septembre 1988: 9; See also the modified proposal of the EEC, *ibid.,* where in an exploratory memorandum it is stated;
   "The modified proposal is drafted in such a way as to guarantee the respect both of rules accepted by or acceptable to our democratic society and of the integrity and dignity of the person, not only while the programme is running but also in its medium- and long-term consequences."

31. Dausset J. *Éditorial: Les droits de l'Homme face à la science. Cahiers du M.U.R.S.* 1989; 3:3, at article X:
   "*Les connaissances scientifiques ne doivent être utilisées que pour servir la dignité, l'intégrité et le devenir de l'Homme. Nul ne peut en entraver l'acquisition.*"
   The Valencia Declaration. Workshop on International Cooperation for the Human Genome Project, Oct. 24-26, 1988, which recognized the responsibility of scientists to ensure that genetic information be used only to enhance human dignity (Roberts L. Carving Up the Human Genome. *Science* 1988; 242:1244 at 1246). See also, Knoppers BM. *L'élaboration d'un code de conduite international en matière des technologies de reproduction.* Report

and Resolutions of the International Law Association conference held in Warsaw, Poland, 1988; London: Int'l Law Ass, 1988, 879.

32. See the explanatory memorandum of the proposal of the Commission of the European Communities, *supra,* note 29 and the citation in note 30. This modified proposal while mandating discussion on the ethical, social and legal aspects of human genome analysis (sec. 4.4) states in section 4.4.4.:
"That the development and the application of somatic gene therapy are not provided for within the framework of the present programme". [Furthermore,] "all contracting parties must abstain from all research seeking to modify the genetic constitution of human beings by alteration of germ cells or of any stage of embryo development which may make these alterations hereditary..."

33. Plomin R. The role of inheritance in behavior. *Science* 1990; 248:183:
"Quantitative genetic research will be important in this endeavor in order to assess the extent to which genetic variance accounts for phenotypic variance and the extent to which individual genes account for genetic variance." (at 187).

34. Kimura M. *Théorie neutraliste de l'évolution.* Paris, Flammarion, 1983.

35. Clarke JTR, Gravel RA, Mahuran DG. *Carrier screening for Tay-Sachs disease: Prospects for direct mutation analysis.* In: BM Knoppers & CM Laberge (eds), *supra,* note 4, 179.

36. Harper PS. The muscular dystrophies. Chapter 118. In: Scriver CR, Beaudet AL, Sly WS, Valle D, Stanbury JB, Wyngaarden JB, Frederickson DS (eds) *The Metabolic Basis of Inherited Disease,* 6th ed. McGraw-Hill, New York, 1989:2869.

37. Evans RM. The steroid and thyroid hormone receptor superfamily. *Science* 1988; 240:889.

38. Ponder B. Gene losses in human tumours. *Nature* 1988; 335:400; Reik W, Surani MA. News and Views: Genomic imprinting and embryonal tumours. *Nature* 1989; 338:112; Editorial. Origins of genetic disease. *Lancet* 1990; 1:887-8, where it is stated:
"Passage of certain genes through male rather than female gametogenesis can determine whether they are to be expressed after fertilization and indeed throughout the subsequent lifetime of the offspring. This process of "imprinting" implies some differences in the condition of the DNA in sperm and egg without violating the principle that the actual base sequences (the genetic code) remain the same'".

39. Rowland LP. Dystrophin: A triumph of reverse genetics and the end of the beginning. *New Engl J Med* 1988; 318:1392; Mandel JL. Dystrophin: the gene and its product. *Nature* 1989; 339:584; Friedman T. Opinion: The human genome project — Some implications of extensive "reverse genetic" medicine. *Am J Hum Genet* 1990; 46:407. See also Goodfellow PN., *supra,* note 17.

40. Scriver CR, Laberge C, Clow CL, Fraser FC. Genetics and medicine: An evolving relationship. *Science* 1978;200:946.

41. Plomin R., *supra,* note 33.

42. See generally, D Weatherall & JH Shelly (eds.) *Social Consequences of Genetic Engineering.* Excerpta Medica, Int'l Congress Series. Amsterdam: Elsevier, 1989; The Hague: Netherlands Scientific Council. *The Social Consequences of Genetic Testing,* 1990. See also Niermeyer MF. Genetic screening and counselling — Implications of the DNA technology. In: Z Bankowski & JH Bryant (eds) *Health Policy, Ethics and Human Values: European and North American Perspectives.* Geneva: Council for International Organizations of Medical Sciences, 1988, 77, at 87.

43. Scriver CR. Human genes: Determinants of sick populations and sick patients. *Canad J Pub Health* 1988; 79:222.

44. Chapple JC, Dale R, Evans BG. (Screening for disease) The new genetics: Will it pay its way? *Lancet* 1987; 1:1189-91.

45. Holtzman NA. Proceed with Caution. *Predicting Genetic Risk in the Recombinant DNA Era.* Baltimore: Johns Hopkins University Press, 1989.

46. McGourty C. News: Human Genome: Public debates on ethics. *Nature* 1989; 342:603. Laberge C and Knoppers BM: *Genetic Screening: From Newborns to DNA Typing.* In: BM Knoppers & CM Laberge (eds), *supra,* note 4, 379.

47. *Supra,* note 14. See also De Wachter MAM. *Screening and counselling — Ethical and policy aspects.* In: Z Bankowski & JH Bryant, *supra,* note 42, at 72.

48. *Supra,* note 13.

49. The term was coined by Capron A. Unsplicing the gordian knot: Legal and ethical issues in the 'new genetics'. In: A. Milunsky & G. Annas, (eds), *Genetics and the Law III.* New York: Plenum, 1985, at 23-4.
50. For an excellent study on genetic information as a "liberty interest", see: Poullet Y. *Le fondement du droit à la protection des données nominatives: propriétés ou libertés?* In: E. MacKaay (ed) *Nouvelles technologies et propriété,* Université de Montréal: Centre de recherche en droit public, 1990 (To be published). See also Macklin R. *Mapping the human genome: Problems of privacy and free choice.* In: A Milunsky & GJ Annas (eds), *supra,* note 49, at 107 for discussion of the public health model versus the clinical practice model in the context of human genetics.
51. Knoppers BM, Laberge C: DNA sampling and informed consent. *Canad Med Ass J* 1989; 140:1023-28.
52. *Supra,* note 13.
53. Andrews LB. *DNA testing, banking and individual rights.* In: BM Knoppers & CM Laberge (eds), *supra,* note 4, 217.
54. Capron A. Decoding the human genome: Dominion, deontology and deconstruction. In: International Conference on Bioethics, *supra,* note 1, at 187. See also De Wachter, *supra,* note 47, at p. 71: "... genetic medicine is turning the winds of individualism in more altruistic directions", and at p. 72: "... Some already say that genetic information is the common property of the family as a 'corporate personality'." See also DC Wertz & JC Fletcher (eds). *Ethics and Human Genetics: A Cross-cultural Perspective.* New York: Springer-Verlag, 1989.
55. *Supra,* note 48.
56. Knoppers BM. *L'appropriation du vivant: Le matériel génétique.* In: E. MacKaay (ed), *supra,* note 50 and Knoppers BM and Laberge CM, *supra,* note 51.
57. Galloux JC. *L'impérialisme du brevet.* In: E. MacKaay (ed), *ibid.*.
58. Fishman R. Patenting human beings: Do sub-human creatures deserve constitutional protection? *Am J Law Med* 1989; 15:461.
59. See generally, Holtzman, *supra,* note 12 and also de Wachter, *supra,* note 47.
60. See Knoppers, *supra,* note 14.
    J.J. Rousseau, *Du contrat social.* (Paris: Pléiade, t. III, 1985), at 347 parag. 361:
    "Chacun de nous met en commun sa personne et toute sa puissance sous la suprême direction de la volonté générale et nous recevons en corps chaque membre comme partie indivisible du tout..."
61. J. Rawls, "Justice as Fairness: Political Not Metaphysical" (1987) *Phil and Pub Affairs,* 223. In this article at footnote 20, at 237, Rawls explains that it was an error in his seminal work, *A Theory of Justice* (Cambridge, Mass: Harvard Univ. Press, 1971), "to describe a theory of justice as part of the theory of rational choice...".
    "What I should have said is that the conception of justice as fairness uses an account of rational choice subject to reasonable conditions to characterize the deliberations".
    Ibid. For an excellent critique of Rawls, see C. Audard et al, éd., *Individu et justice sociale: autour de John Rawls,* (Paris: Ed. du Seuil. 1988). See also M.A. Hermitte, "Le droit civil du contrat d'expérimentation," dans Fondation Marangopoulos pour les droits de l'homme, *Expérimentation biomédicale et Droits de l'Homme,* à la p. 39:
    "Les tendances actuelles de l'expérimentation sur l'homme répondent donc assez exactement à la philosophie de John Rawls, mélange d'utilitarisme et de respect des droit de l'homme; l'utilitarisme fournit la base de raisonnements que la philosophie des droits de l'homme et du contrat social vient réorganiser plus ou moins profondément".

PUBLIC POLICY IMPLICATIONS OF THE HUMAN GENOME PROJECT

Robert Mullan Cook-Deegan*

The human genome project will deeply affect many aspects of science and science policy, and will ramify into social policies related to the use of genetic tests and exchange of genetic information about individuals. I shall focus on only a few among dozens of policy issues. I shall first compare funding the genome project to the funding of other biomedical research, an issue that continues to dominate the debate within the scientific community. Several other science policy issues arise from the larger scale of genome research. One concerns difficulties related to research involving large pedigrees. The other stems from complications of information exchange, sharing of research materials (such as clones, probes and cell lines) and how to resolve conflicts that arise when more than one research group wishes to pursue similar work. These issues will be difficult even within national research establishments, and even more difficult when international scientific collaboration is involved.

I shall then turn to public policies related to the use of genetic information, rather than conduct or administration of science. I shall begin with a brief discussion of issues related to the use of genetic test results by insurers and employers, focusing on the issues likely to arise first in legislatures and courts. I shall finally turn to long-term issues. The responsibility of government to promote collective rather than individual goals has been used as an argument for state intrusion into reproductive decisions. I shall specifically discuss whether there are state interests in protecting the "human gene pool" that can override individual or parental decisions.

Science Policy Issues

Funding the Genome Project

Scientists expressed concern about whether the genome project would detract from or displace other biomedical research from the beginning, at meetings in Santa Cruz [June 1985], Santa Fe [March 1986 and January 1987] and Cold Spring Harbor [June 1986][1]. The earliest memos in the National Institutes of Health (NIH) and the Department of Energy feature this issue prominently, and it has persisted as the principal source of controversy within the biomedical research community.

Arguments about displacement of other research fall into two basic groups. One group contends that the genome project is a vehicle for attracting "new" funding into biomedical research, because of its direct appeal

* National Center for Human Genome Research, National Institutes of Health, Bethesda, MD, U.S.A.

to an objective that the general public and Congress can readily understand. Against this view is the contention that a massive concerted effort will redirect funds from biomedical research, particularly at NIH, because it sucks funds away from existing programmes and diverts it to more targeted research.

In my view, there is a seed of truth in both arguments, but both reveal a simplistic notion of how science policy decisions are made. First, both NIH and the Department of Energy indeed took at least some of their first funds for the genome project from existing budgets not initially programmed for genome research. The 1987 funding of $4.5 million at the Department of Energy was reprogrammed from previously appropriated funds, and NIH staff estimate that roughly $5 million of the original $17.3 million given to the National Institute of General Medical Sciences would have been used for other purposes. The initial NIH funding constituted just under 2% of a supplement appropriated by Congress, above the President's request for fiscal year 1988.[1] Would this initial funding have gone to investigator-initiated (R01) grants if there had been no genome project? Since these funds can be traced to NIH Director Wyngaarden's reply to questions from the committee, any funds would have been reallocated according to NIH's own budget priorities. These funds might not have been allocated at all, or they might have been allocated to NIH. It is impossible to judge this in retrospect. If funds had gone to NIH, some would have gone to investigator-initiated research, clinical trials, creation of new centres, animal care costs and other special initiatives, in rough proportion to the overall NIH budget. It is at least clear that there was no direct robbing of Peter to pay Paul within the NIH budget. Those asserting either extreme position, that funds were entirely "new" just because of the genome project's attractions or that they were carved out of R01 funding, are on shaky ground.

The National Research Council panel made its recommendation of a $200 million budget contingent on some notion that the funds would be incremental, and not taken from existing programmes[2]. But this is a criterion impossible to meet, since the budget process does not consider line items one at a time, but rather as part of a package presented by agencies. The question then becomes: "Should the NIH have recommended funding for the genome projects?" This is a more meaningful approach, and makes it immediately apparent that the attempt to isolate "new" money cannot truly side-step the need to compare genome research to other priorities in the same budget.

Is the genome project worth 2 to 3% of the overall NIH budget? My own answer is yes, because the collective nature of generating maps and sequence data, coupled with an explicit recognition of technology development and automation, fills gaps left unfilled by previous NIH research programmes. Consensus on the merits of a genome programme followed the history of difficulty in systematically developing a genetic linkage marker map, in mounting physical mapping efforts of human chromosomes, and in dedicated efforts to develop new technologies and analytical instruments. The debate about the genome project highlighted the neglect of these and other collective resources, in stark contrast to NIH's unsurpassed management of most biomedical research. There are many reasons for this weakness, including how such proposals fare in peer review, how applications under the Small Business Innovation Research programme must be targeted, and the inability of NIH to fill gaps when investigators in the community fail to do so through the normal grant process. Similar issues would doubtless have surfaced regarding physical mapping of human chromosomes and projects encompassing the entire genomes of other organisms. NIH has played a role in developing computer analysis, automation, instrumentation and other kinds of technology development, but there have been some significant blind spots. No NIH funds were used for the early work on the Caltech DNA sequenator, for example. The Weingart Institute, Upjohn, Monsanto and the Baxter Foundation funded this work until there was a working prototype. When government funding commenced in 1984, it came from the National Science Foundation, not NIH.

Whatever the reasons, there was clearly an institutional problem in mounting a concerted research effort on the scale of the human genome. Opinion leaders of molecular biology recognized this. The National Research Council Committee forcefully articulated a need for such large-scale collective efforts. The question is indeed precisely whether we are better off with a genome project of a few hundred million dollars, or another 2 to 3% of NIH funding for other purposes. I have little doubt that a systematic effort is vastly less wasteful in the long run. The point is thus not whether the funds were "new", but whether there is merit to collective map-making and technology development. Should molecular biologists be encouraged to beat their separate individual paths into the wilderness, or should they build a few roads that will make the overall process faster and cheaper in the long run? The answer is clear, although it does not permit a quantitative answer of exactly how much is the right amount of road-building as distinct from undirected exploration. A few per cent seems about right.

Large pedigree research

Maps of the human genome will be constructed from a variety of DNA sources — some from families, some from cell lines and some from individuals. However maps are constructed, their use and interpretation will inevitably involve large numbers of people in pedigrees.

This area has been left to individual investigative groups and collaborations to deal with as they see fit. This is understandable and appropriate, but the increasing scale of research and increasing availability of technologies mean that more investigators can use them, and may well increase demands on large families afflicted with genetic diseases. The time may be near when these issues demand a more consistent policy. The search for genes associated with Huntington's disease, cystic fibrosis, Alzheimer's disease, colonic polyposis and Tourette syndrome has involved many pedigrees, some of them quite large. Each research group or collaboration has somewhat different policies. Some groups as a matter of policy do not even note in the medical record that a genetic test has been done, for fear that insurers and employers who gain access to the record would use it for purposes of discrimination. Some groups segregate the genetic records from the rest of the medical file. Some groups publish false pedigrees to avoid inadvertent disclosure and invasion of privacy within large families (to prevent a family member reading a scientific article and learning medical facts about others in the family, for example).

These policies are not, however, widely discussed or even known throughout the human genetics community. The *Centre d'Etude du Polymorphisme Humain* (CEPH) had to develop some policies regarding exchange of clinical information and patient identifiers in putting together its genetic linkage mapping collaboration[3]. The Huntington's disease collaboration has a process for gaining access to its University of Indiana data-base, managed by P. Michael Conneally. The University of Utah (the largest contributor of family pedigrees for genetic linkage analysis) also has specific policies to protect the confidentiality of families used in the CEPH collaboration, and also governing scientists' access to the enormous Mormon genealogical data-bases stored near Salt Lake City.

These *ad hoc* arrangements address the specific needs of particular scientific groups. The clinical and genetic aspects of several diseases vary greatly, so there will never be one standard process for protecting the confidentiality of genetic data in large families. The time may soon be ripe, however, to consider the issues more systematically, since so many more genetic markers will become available, and many more groups will then have the tools to do genetic linkage and to study human genetic disorders. This suggests that an increase in scale and broadening of the base of investigators may well cause problems. It is all too easy to imagine how very private information about some disorders could harm individuals, particularly data related to behavioral and cognitive characters. This includes many fairly common conditions, such as bipolar disorder, Huntington's disease, Alzheimer's disease, the Fragile X syndrome, schizophrenia and Tourette syndrome.

I have no solutions to offer. I predict only that there will be a need to address these issues explicitly over the next decade. Several important but incompatible benefits must be weighed — free exchange of data and materials, protection of patient privacy, rigorous clinical assessment of patients entered into pedigrees, integrity of the scientific literature, and needs of family members who seek to ascertain information about their relatives.

Sharing of data and materials

The collective nature of genome research is a distinctive characteristic. Most scientific work results in facts published in the literature and slowly integrated into the fabric of science. Map-making and sequencing add some additional complications, since it is often necessary to compare the results of different groups, using the same materials. For example, it is impossible to order genetic linkage markers relative to one another without directly testing the probes against one another in a common set of meiotic cells. This is the basis for the CEPH collaboration and is a feature of other genetic disease searches. The issue is thus not new, only more complicated, because of an increase in scale and a sense that the stakes are high. Career aspirations and proprietary concerns may come into conflict with incentives to share data. Hunting for disease genes is a probabilistic process, much like other areas of science, but the number of individuals involved may be substantially greater. Dozens of groups cooperated in searching for the gene causing Duchenne and Becker muscular dystrophy, and again for cystic fibrosis. One problem with this has been the political (with a small "p") problem that this results in papers with dozens of authors, only one of whom gets primary credit. This has induced the Huntington's collaboration to agree that the paper about finding the gene, when it is finally found, be authored by the collaboration collectively, rather than by a set of individuals ranked by level and relevance of effort. This may or may not prove a difficult issue when investigators, particularly assistant professors and post-doctoral workers, come up for tenure review, academic promotion and grant review. If they have no publications that clearly specify their contributions, or if they act as individuals in a large collective effort, will they fare well when measured by those who have pursued smaller and more traditional questions in science? Physics, chemistry and multi-centre drug trials have had to contend with this issue, and now it is happening in molecular biology.

The issue obviously surfaced even before the genome project made any impact on human genetics. The Duchenne and Huntington collaborations, for example, were formed before there was a genome project. The genome project, with its emphasis on generating information for maps and sequence data-bases, will nonetheless accentuate the collective nature of human genetic research.

Exchange of materials and information deemed commercially useful may emerge as another cause of friction. Policy-makers increasingly view public science funding as a national investment. When seen in this light, research results with commercial potential should be managed to secure the maximum advantage to the nation whose taxpayers funded the research. The US Congress has sent several clear signals that publicly funded science should be commercialized as much as possible. Several new patent laws, executive orders and technology-transfer statutes from the 1980s decentralized intellectual property rights from government to research institutions, with the intent of providing incentives for commercial exploitation.

The map and sequence data derived from public funding should be made readily available, as long as disclosure does not endanger patents. Asking for more openness seems hopeless; asking for less seems irresponsible. Maps and sequence data-bases will be useful only to the extent that they are complete. If GenBank and EMBL Bank contain only a small fraction of sequences, and if many data are held privately, then comparing a new sequence to the data-base will be far less useful. If maps show only a minority of markers, and those markers are not readily available for locating genes, then the maps are relatively useless. In most cases, there will be no direct conflict between exchange of information and commercial interests. There will be a gray zone, however, when investigators believe that map and sequence data need to be held long enough to file a patent based in part on such data. The extent of that gray zone is unclear. It should be kept as narrow in time and scope as possible.

Much sharing takes place normally in the scientific process, of course. Human genetics, however, does not have a universally laudable record of sharing. Many who move to human genetics, from yeast or nematode biology, for example, note that the higher stakes, greater edge to competitiveness and larger scale all complicate the scientific norm of free exchange of data. The search for the Down's syndrome and Alzheimer's disease genes on chromosome 21 might well have induced a lapse into open warfare. Quite the opposite has occurred, however, as chromosome-21 researchers recently agreed to share materials and information, and began to establish a mechanism to screen probes.[4] This can perhaps serve as a model for future efforts.

Research targeting

The need for concerted collective production of worldwide genetic maps will highlight a need for planning to reduce duplication, ensure complete coverage, and attend to quality control. There is no sense in having several independent groups produce maps of chromosomes 7, 4, 21 and X, while none work on others (e.g., chromosome 13, 18 or Y). Left to itself, the community would naturally follow the past pattern of shark-like aggregations swimming around the identifiable tasty morsels in the genome. Competitive races to discover the genes associated with important diseases may excite a feeding frenzy as discovery of a gene draws nigh, with neglect of wide genomic expanses. The intensity of competition is not entirely bad, indeed it is good — it is prudent to focus most where returns are most immediate — but the neglect of other areas is not good. How can genes mapping to these areas be located if no one develops the markers? The US effort recently took steps to ensure that at least one group takes principal responsibility for the genetic linkage map of each chromosome. Prominent genetic-linkage mapping groups were invited to get together. Like those working on chromosome 21, they agreed among themselves on how to proceed. In a sense, this constitutes research targeting, since the coordinators for each chromosome agreed to find markers if there were no other road to progress. It was in the long tradition of scientific collaboration rather than coercion, however, since the scientific community decided the issue.

The issues of collaboration and data sharing will surface time and again in the construction of physical maps, where overlapping efforts are even more likely to be wasteful in the long run than in genetic linkage mapping. For now, the technologies to make clone libraries and to order them are sufficiently unsettled that redundancy of effort is desirable, but as the technologies mature the field will have to adapt. It may become necessary to be more systematic in funding grants, taking into account not only the methods used and the richness of innovative ideas, but also the target of the research and the extent to which it overlaps with ongoing efforts.

Social Policy Issues

Genome research will impinge on many social policies. As new knowledge emerges, it will doubtless change how we conceive of genetic disease and genetic variation, and new conceptions will in turn slowly alter policies. For example, recognition that genetic factors influence obesity, alcoholism and many behavioural disorders may alter how we deal with those problems. In the future, discriminating against those who are overweight, now tolerated, may be regarded as barbaric in the same way as racial discrimination. The criminal law may have to adjust to new notions of responsibility. I shall not attempt to cover the whole territory in my brief remarks, but rather confine myself to the three issues that I believe are most pressing: genetic discrimination in private insurance and employment, delivery of genetic services, and dangers of state action to protect the collective "gene pool."

Genetic discrimination

This topic overlaps extensively with issues covered by Dr Knoppers elsewhere in this volume, and therefore I shall focus on presenting the options for public policy. I shall note only in passing how ethical analysis might help to select among the options. There are several pressing policy dilemmas that stem from the same root — a conflict between the increasing power of new technologies to detect genetic heterogeneity and the presumptions underlying private insurance and private employment (specifically, how genetic data are used in underwriting practices). These issues have been discussed in several recent books.[5,6,7] In the discussion that follows, I draw on these books as well as on background papers prepared for the Office of Technology for its 1983 report on genetic testing in the work-place,[8] a 1988 report on medical testing and health insurance,[9] and a report on a work-place genetic testing study now in preparation.

New genetic knowledge and genetic tests will emerge from genome research. Tests will be made faster, cheaper and more accurate, and this is all to the good for medical diagnosis. When such medical information is used in the nonmedical context, however, conflicts soon arise. There is a conflict between standard practices in private insurance and private employment, on the one hand, and the intuition that we should not be treated differently on the basis of genetic characters, on the other. This intuition is based, at least in part, on equating "treated differently" with

62

"treated unfairly," when a person cannot choose the genes he or she was born with, and can do little about them. It is thus a claim of social justice premised on a choice-based conception of fairness.

There are, however, problems with both parts of this equation. Certain practices that are unfair by the same criteria are already used in current underwriting practice. A person with very high blood pressure or very high cholesterol level has equally little control over his or her genes, and yet may be denied life insurance or disability insurance. As these factors can be tested for, and are genetically determined at least in part, this is not a novel issue. A member of a family with Huntington's disease can be denied life insurance or private health insurance, on the basis of answering questions about family background. Also, the fact that the tests to identify risk factors are based on DNA analysis does not mean that they are testing for genetic characters. Underwriters seek information that has strong predictive value, whether genetic or other. Most genetic tests will not have strong predictive power. One must predict how likely a person is to die young (for life insurance), to become disabled (for disability insurance) or to consume health care (for health insurance). Predictive value is thus a combination of several factors: how well a test measures increased morbidity or mortality, severity of effect, and costs associated with the effects. For most genetic tests the predictive value will be low. For certain genetic diseases it might be high, such as cystic fibrosis, a few varieties of hemoglobin disorders, Huntington's disease, etc. For many, it will not be high even if the test is highly accurate, because the clinical variation is so great that the resource-consumption curves overlap extensively with the average population.

It is clear, however, that at least some tests will be powerful tools for underwriting. These tests do pose true public-policy dilemmas. Once such a test is available to the patient, private insurers will need access to the same information to protect themselves against adverse selection. Those who learn they are at great risk have good reasons to buy as much insurance as they can. A few such individuals in the risk pool can distort actuarial tables and put the insurer at financial risk. If they cannot be directly protected from individual "gaming," insurers will at least want a common set of rules so that genetic discrimination does not become a potential basis for competition among private insurers. This means new regulations or new laws.

One public policy option would be to let insurance markets operate as they currently do, and have them slowly adjust to the new technologies, with a few people here and there cut out of the private insurance markets. The problem with this option is that both insurance and employment contain a large dose of social welfare policy. Most of the population presumes that health care and employment are entitlements, at least in part — everyone who wants them should have them. There may be some disagreement about life insurance, but it too has a social-welfare element. These private markets are not purely private enterprises: they serve public purposes, and are

63

therefore regulated. There are several policy options regarding genetic discrimination.

Employers and insurers could be left to decide whom they hire and whom they insure, without further action regarding genetic tests. Under this scenario, insurers would choose whether or not to use any new tests that will be developed. Insurers may find that the new tests do not give them enough valuable information to alter underwriting practices, or they may find that some do, and they will either increase premiums or deny eligibility to some people. They may choose not to use such underwriting information because using it would cause too much of a public outcry, or would call the entire industry into question. Underwriters do not use race as a discriminating factor, for example. They explain this as a policy because it is not a "biological" determinant that is being measured, but rather unfair intervening social factors, such as poverty and unemployment, which influence morbidity and mortality. If this argument were consistently followed, many current practices would likewise have to be abandoned. Are smoking, scuba-diving, and other factors now used in underwriting, biological or social variables? Are differential rates based on gender due to biological or to social differences? It seems that a *moral* judgment is at work as well, based on a tacit agreement that race *should* not be considered, and hence actuarial tables are made to adjust to a policy decision.

The same policy interests could be applied to genetic factors, resulting in a proscription of genetic testing, but this would cause problems. Public policies for insurance and employment pursue different paths. I shall briefly discuss insurance markets and then turn to employment.

We shall probably learn that most of the factors used in underwriting are likely to be at least influenced by genetic factors, so that banning genetic discrimination would be tantamount to banning all underwriting, essential to private insurance. Could government mandate only that DNA-based tests be proscribed? This would be unfair to those groups for which a protein-based test of a genetic character could be developed. It would not get around the basic problem. Proscribing DNA testing in underwriting would also complicate existing practices, since, more and more, laboratory tests are DNA-based, including those testing for infections by HIV and other infectious agents. There might be room for proscription of those factors deemed uncontrollable by the individual, at least as a guiding principle, but this would leave a large zone of uncertainty when genetic factors are uncertain and the issue of individual control is unresolved. The dilemmas likely to arise in connection with alcoholism, hypertension and cancer are obvious sources of dispute along these lines.

It is difficult to predict what will happen, but it is worth pointing out that this is not an entirely new issue. The balancing act between narrowing the population eligible for insurance and containing insurers' financial risks is the very meat of underwriting, and has always been a problem. Insurers have little interest in getting such refined predictors of disability and mortality that they can sell insurance to only a small fraction of the general population. This defeats the purpose of insurance, dramatically increases the transaction costs of marketing products and assessing new clients, and narrows the potential market. Private citizens have little to gain by entirely dismantling a system that works for many people, It is, of course, possible that the public will suddenly awaken to the fact that private insurance markets are unfair to some individuals and will rise up to make employment and health insurance — perhaps even life and disability insurance — public entitlements. This could be done by public subsidy of private insurance, by public mandate, or by direct public assistance programmes. Public subsidy could entail tax incentives or direct payments. Congress or state legislatures could mandate coverage in much the same way that automobile insurance is a private enterprise, but is required in order to drive in many states and countries. A public insurance scheme could involve either a pool for those ineligible for private insurance or mandatory national insurance programmes such as national health plans. Increased knowledge about genetic heterogeneity thus does pose an unprecedented and insurmountable dilemma, but, as in so many other policy areas, the inherent inconsistencies can be patched over by compromise and satisficing, if consumers and insurers are willing to negotiate a new social contract.

Job discrimination is likely to be applicable to a much narrower range of disabilities than is insurance, as most employers would likely worry about only those disabilities that might affect job performance. Protection from genetic discrimination can be regarded as a civil right, proscribing discrimination unless an employer can argue that it is in the public interest or key to an employer's job-related interests. Public interests can be asserted either because genetic risks impose undue costs, subject the person to undue risk, or endanger public safety. It might be possible for airlines to ask for more frequent check-ups among those with hyperlipidemia or carrying the Huntington gene, for example, but employees could not be fired unless they were shown to be disabled. It would not be possible to fire at will on the basis solely of genetic testing. In the United States, passage of the Americans with Disabilities Act has extended this kind of protection to most employers and employees, although the scope and degree of protection will have to be worked out through regulations and case-law. The statute itself is silent on the use of genetic tests, but most legal scholars agree that those with genetic disabilities would certainly qualify, and even those "considered to be disabled" are covered. It is not entirely clear how the statute applies to those not disabled but *predicted to become* disabled — for example, after a genetic test — but it seems likely that courts will interpret the statute to cover such individuals. Having a law, of course, does not solve the problem but it does go some way towards doing so. There may be further legal and statutory initiatives in the coming years. Most likely, these will centre on the confidentiality of genetic information and the voluntariness of seeking genetic tests.

Despite some legal protection, there may still be a long-term problem in employment discrimination. In the United States, private insurance and employment are closely linked. Health insurance and life insurance are part of an employment package for most people. This is particularly true when employers choose to handle their own health plans, through self-insurance, rather than through a third-party, private insurer. More and more, employers are following this path in order to gain direct control over rising health costs. Employers thus have incentives not to hire those predicted to incur disproportionate health costs before retirement. In this case, policies regarding employment and private insurance converge to the same sets of options discussed above under insurance.

Delivery of genetic services

Dorothy Wertz's and John Fletcher's recent survey of clinical geneticists in 19 nations[10] showed many differences in how genetic counsellors handled different clinical scenarios, but one issue was identified as a future problem in almost every country: the difficulty of meeting demand for services. As the conceptual foundations of medicine have shifted towards genetics, health-care resources have not kept pace. Several studies show that genetic services, particularly the time-consuming counselling components, do not pay for themselves financially.[11,12,13] While genetic tests will be less costly and more readily available, it is not clear that the necessary ancillary services will change similarly. The likely outcome is greater use of tests, with a smaller proportion of those tested having counselling and education to understand the results. This is an issue not only for health professionals who order the tests, but also for families who want the information. Most genetic services are delivered in conjunction with academic health centres, for a variety of historical, financial and logistic reasons. As genetic tests become more plentiful, useful and inexpensive, more will be used. More important, as new tests are developed for disorders prevalent in the population, such as cystic fibrosis and the muscular dystrophies, it will become necessary to design services accessible to a much broader public. Italy and Greece have already faced issues of population screening because of their high prevalence of hemoglobin disorders (see the Cao chapter in this volume). Testing for cystic fibrosis will soon become a policy issue in other nations with mainly Caucasian populations, including most of Europe, Latin America and North America. Eventually, genetic services must be integrated into the mainstream of medical care. This means that primary-care professionals must learn more about genetics, and will likely even provide some care directly — perhaps even fully managing the more common genetic disorders.

The emergence of human genetics will require several policy responses. First, health professionals have to learn about genetics and genetic disorders. This will include dramatic improvements in nursing schools, psychology programmes and medical schools, but also continuing education for the much larger group of health professionals already in practice. Genetics can no longer remain the exclusive province of highly specialized tertiary — and quaternary — care centres. There will always be a need for specialized genetics clinics, as primary-care professionals will never learn about the thousands of rare genetic syndromes, but the existing model cannot be expanded sufficiently to meet the likely future demand. Specialization will continue, but medical practice more generally must adapt; perhaps medical genetics will grow from its current state to a robust and large medical speciality. Whether it does or not, it is clear that mainstream medicine will have to deal with more genetics in the future than it has dealt with in the past.

Second, the public must become much more sophisticated about genetics. Most people in the general public will learn about genetics by learning about, and worrying about, new developments reported in the press, or when the new technologies directly impinge on their lives. But there is as well a responsibility to educate young people so that they may understand the world in which they live, and this will take a concerted effort on the part of scientists and health professionals in the field.

Health care will also have to adjust to the new knowledge. The current system was built to cure acute diseases, but this is now only a minority component of health care. Long-term incurable disabilities that require sustained medical management now fill the days of doctors and nurses. Genetic diseases will mainly fall into the category of chronic illness, with the further complication that few people understand the basics of heredity and molecular biology. Genetic counselling is largely devoted to education about genetics and disease. This is the main cost component, and the principal source of confusion among those counselled. This will remain true for the foreseeable future. In 20 years, perhaps genetics will be so much a part of our daily lives that everyone will know at least something about it. That may be a vain hope, but it is possible.

Meantime, the health care system will have to provide genetic education and counselling to a larger fraction of the population, about a greater variety of disorders and a more diverse range of tests. This will require more resources, and securing those resources will likely take considerable patience and political skill. In nations with national health schemes run by government, policy-makers will act only when persuaded of the need to reallocate resources, in competition with other pressing concerns. In countries with private payers, clinical geneticists will have to contend with a bias towards procedure-oriented reimbursement methods. Payment methods are being devised to attempt to reward direct patient contact, or "cognitive services" demanded in genetic diagnosis and counselling, but given the history of health policy initiatives there is room for skepticism that the new means will reach the intended ends. I wish that I could feel sanguine about the prospects, but I foresee a painful process. By then, the genetic paradigm will be so dominant that it will be an impediment to the next wave in the continual revolution of medical knowledge.

Is there a State interest in protecting the gene pool?

Two answers to this question seem appropriate to me: "Only rarely" and "Why do you want to know?" This is an area where public policy must proceed with caution, in Dr Holtzman's apt phrase[5]. The obvious reason for concern is the repugnant history of eugenics, followed by racial hygiene, in the first half of this century. Recent scholarship shows that eugenics was not an ideology imposed on science by politicians (as one might argue was the case of Lysenkoism), but rather an ideology developed first among reputable scientists and doctors and exported into the political realm with their full blessing.[14-17] Galton and Fisher are still founders of some parts of human genetics, even if Davenport and Lenz do not fare as well in the history of science. The giants of genetics who promoted eugenics left intellectual legacies well beyond their policy prescriptions, but their policy prescriptions were awful indeed.

Eugenics rested on the premise that genes matter a lot, and that the way to change their distribution in the population was to change reproductive behaviour — specifically, to limit options for prospective parents. One could encourage those with good genes to reproduce disproportionately, or one could reduce reproduction among those with bad genes. Since policy carrots, or noncoercive encouragement, were expensive, the positive options were never fully implemented. Policies to limit options among the undesirables were relatively more attractive, in part because of lower direct costs, and presumably also because they found fertile soil in latent racism. There were sterilization statutes to prevent reproduction among the retarded and behaviorally aberrant, enforced mainly in California and Virginia and then in Germany.[17,18] Where eugenics was linked to racial hygiene, immigration statutes and antimiscegenation laws were a natural consequence. The logical extreme was to obliterate those carrying undesirable genes from the population so that they could not reproduce - by genocide.[15,16,17]

There was a presumption that the greater good of the greater number justified constraining individual liberty. Of course, this is not the only justification in the chain of arguments that led to the Holocaust, but it is was an important one, and an especially questionable one, when the greater good is preservation of the gene pool. The arguments of the eugenicists are technically sound. It is perfectly clear that if one preserves bad genes for many generations they become more prevalent, and that if good genes are selected they become more plentiful in the population. The problem is not the arithmetic, but rather the time-scale and the difficulties in judging good and bad.

Fisher was an unrepentant eugenicist and racial hygienist, but his own calculations suggested that it would take several hundred generations to alter the prevalence of a rare recessive character.[19] It is very difficult to alter the prevalence of a character with a high mutation rate, such as Duchenne muscular dystrophy. These arguments do not apply to dominant traits, like Huntington's disease (which could theoretically be almost eliminated in a single generation), but Huntington's is the exceptional case. It is difficult to imagine seriously justifying a public policy when its effects will require persistence on a time-scale roughly equal to the entire history of civilization. There are few policies that have been pursued with great consistency since the time of Christ.

The eugenicists were perhaps right that characters could be cultivated by creating selective pressure on the population, but they were wrong to think that people would put up with it. They were even farther from the mark in thinking that people *should* put up with it.

Even if the policies could be effective, there are further problems: 1) the selected character must be strongly determined by genetic factors, or altering reproductive options will not work; 2) there must be clarity about what is to be selected for and against; 3) there must be some reason that the State and not others should make the choice; and 4) policies should be the least restrictive available options (i.e., alternatives are unavailable or less acceptable). I shall briefly deal with these in turn.

We have learned about genetic factors, for the most part, by thinking about Huntington's disease, hemoglobin disorders, and other diseases inherited in Mendelian fashion. These are, in the larger framework of all medical genetics, the unusual cases. Genes associated with these diseases obviously have a strong impact. Yet most diseases are not monogenic, and most mutations, even of protein-coding regions, probably do not cause detectable illness. There has to be slop in the genome, or we could not tolerate even the very low background rate of mutation. Even if the mutation rate is one in 100,000,000 per generation, there would still be a few hundred new mutations in the average human genome. The degree to which selection works will depend on how many genes are involved and how they interact. It is not clear that it would be effective to select simultaneously for strength, height, colour perception, drawing skill, musical ability, intelligence and physical appearance. Indeed if the policies preventing reproduction were vigorously enforced, the population explosion would quickly be over, at least in my family.

Even if policies could be sustained for millennia and focused on one or two really important characters, what would those characters be? I truly do not know. The full richness of human existence does not depend entirely on one or two characters. It helps to be smart and it helps to be strong, but our heroes are not so easy to characterize. Einstein was not the smartest scientist of his day, nor Gandhi necessarily the wisest political leader. Franklin Delano Roosevelt was far from the most virtuous and honest American president in modern history, but arguably he accomplished more of lasting value than anyone else. I despair of trying to select for desirable characters systematically and collectively. The obvious danger is that evanescent cultural values will be translated into constraints on reproductive choice, justified by genetics but in fact motivated by parochialism.

The most devastating impacts of eugenics and racial hygiene were their infringements on personal liberty and human rights. The eugenicists did not generally argue for positive selection, or they would quickly have confronted the intellectual difficulties noted above. Rather, they usually pointed to identifiable harms that could be prevented — for example, mentally retarded children born to mentally retarded parents — and asked when it would end. Even Oliver Wendell Holmes, one of America's pre-eminent legal minds, fell into this trap in the infamous mandatory sterilization case,

Buck v. Bell. His decision was not only morally questionable in subsequent analysis, but factually in error. A child of the sterilized woman proved not to be retarded.

The justification of State action hinges on the *parens patrie* doctrine — the State will protect interests that individuals cannot protect for themselves. In the context of reproductive choice, this makes sense only if we are willing to accept that the State can decide better than parents themselves what kind of children they can have. This seems exceedingly dangerous in general, and all the more so in the context of genetic disease. Many families afflicted with genetic disease learn about its effects by direct experience with an affected relative. When this is the case, and particularly when parents have previously borne a child with a particular disease, they have much more relevant information than any professional can ever hope to have. They are certainly better placed to make judgments about future children than a legislative body. The worst case is that prospective parents have more or less the same information available to them as the general public. Of course, the argument in cases of mental retardation is that the subjects have access to information but cannot comprehend it, or they make bad decisions. Some will argue that parenthood should only be conferred on the fully autonomous - requiring intellectual competence, understanding of the choice at hand, and appreciation of its consequences. This argument also has some unfortunate implications.

If humans were fully autonomous when they chose to become parents, then why did nature have to trick them into parenthood by orgasm? Autonomy, as I understand it, is not a binary state but an ideal. It involves competence to make a decision, understanding of the background facts and consequences likely to ensue from different decisions, and freedom from coercive influences. None of us has full understanding or is completely free of influences. If we are to constrain autonomous choice regarding parenthood, we are thus all at risk, although to varying degrees. This is an area where good arguments can be made on both sides, but any coercive policies invoking the power of the State must overcome strong presumptions in favour of parental autonomy. I am willing to accept a few people in the world with characters I consider suboptimal in order to avoid a comprehensive written and oral examination for prospective fathers.

Constraints on reproductive decisions must not only overcome a presumption of parental autonomy, but also be the least restrictive options at hand. In the case of selecting for intelligence, it might be possible to increase the population's problem-solving abilities by imposing selective pressure for several hundred years. We could, for example, kill off or sterilize those who score lower than 70 on an IQ test, and give tax breaks and free access to *in vitro* fertilization for those scoring over 130. The net result might be a marginal increase in overall intelligence over time. (I am willing to grant this only for purposes of argument.) Is this the best way to achieve that end? What could happen if the expense of all those sterilization procedures and free IVF cycles were channelled into prenatal care, prevention of drug abuse and education? There is almost certainly no purely analytical framework adequate to make this calculation, but the latter, social-policy, steps are at least arguably as likely to be effective as the genetic strategy. They are clearly ethically preferable. The positive social measures do not inherently violate autonomy or notions of fairness. To generalize from the discussion of intelligence, let us take strength or height or any other human characteristics we value. These are dramatically improved by nutrition and exercise. Is there any case where a trait is strongly influenced by a gene or two, and for which there is no less intrusive social recourse? There may be, but I have not thought of it.

This concluding section is a long argument for a single position - that parental autonomy should, in general, prevail over notions of collective good. In the genetic context, preferences for the greater good have historically proved to be mere cultural bias masquerading as genetics.

References

1. Cook-Deegan, R.M. *The Human Genome Project: Formation of Federal Policies in the United States, 1986-1990.* Committee to Study Decisionmaking, Institute of Medicine, National Academy of Sciences, Washington, DC. 1990.
2. National Research Council. *Mapping and Sequencing the Human Genome.* National Academy Press, Washington, DC. 1988.
3. Dausset, J. *et al.* Centre d'Etude du Polymorphisme Humain (CEPH): Collaborative Genetic Mapping of the Human Genome. *Genomics* 1990; **6**(3): 575-77.
4. Roberts, L. Genome Project: An Experiment in Sharing. *Science* 1990; **248**: 953.
5. Holtzman, N.A. *Proceed with Caution.* Johns Hopkins University Press, Baltimore, MD. 1989.
6. Nelkin, D.A. & Tancredi, L. *Dangerous Diagnostics: The Social Power of Biological Information.* Basic Books, New York. 1989.
7. Rothstein, M.A. *Medical Screening and the Employee Health Cost Crisis.* Bureau of National Affairs, Washington, DC. 1989.
8. US Congress, Office of Technology Assessment. *The Role of Genetic Testing in the Prevention of Occupational Disease.* Government Printing Office, Washington, DC. 1983.
9. US Congress, Office of Technology Assessment. *Medical Testing and Health Insurance.* Government Printing Office, Washington, DC. 1988.
10. Wertz, D.C. & Fletcher, J.C. Eds. *Ethics and Human Genetics: A Cross-Cultural Perspective.* Springer-Verlag, New York. 1989.
11. Pyeritz, R.E. *et al.* The Economics of Clinical Genetics Services, I. Preview. *Am J Hum Genet* 1987; **41**: 549-58.
12. Bernhardt, B.A. *et al.* The Economics of Clinical Genetics Services. II. A Time Analysis of a Medical Genetics Clinic. *Am J Hum Genet* 1987; **41**: 559-65.
13. Bernhardt, B.A. & Pyeritz, R.E. The Economics of Clinical Genetics Services. III. Cognitive Genetics Services Are Not Self-Supporting. *Am J Hum Genet* 1989; **44**: 288-93.
14. Kevles, D.J. *In the Name of Eugenics.* University of California Press, Berkeley, CA. 1985.
15. Lifton, R.J. *The Nazi Doctors.* Basic Books, New York. 1986.
16. Muller-Hill, B. *Murderous Science.* Oxford University Press, New York. 1988.
17. Proctor, R. *Racial Hygiene.* Harvard University Press, Cambridge, MA. 1988.
18. Reilly, P. Genetic Screening Legislation. *Adv Hum Genet* 1975; **5**:319-76.
19. Fisher, R.A. *The Genetical Theory of Natural Selection* (Second Revised ed.). Dover, New York. 1958.

ANTENATAL DIAGNOSIS OF β-THALASSEMIA IN SARDINIA

Antonio Cao[*]

Introduction

The β-thalassemias are a group of genetic disorders occurring with a high frequency in the Sardinian population. The incidence of the homozygous state is 1:250 live births, the carrier rate is 1:8 persons, and 1 couple in 60 is composed of two β-thalassemia carriers and thus at risk of producing the homozygous state known as thalassemia major.[1] Thalassemia major is a severe anemia, which, without treatment, leads to death within the first decade. Modern treatment with continuous transfusions and iron chelation by daily subcutaneous infusion with desferrioxamine may permit a long survival. However, desferrioxamine treatment is troublesome and associated with low compliance. Bone-marrow transplantation from HLA identical donors may be an alternative,[2-3] but is still associated with a high rates of mortality and morbidity.

The high frequency and severity of thalassemia major in the Sardinian population, and the availability of procedures for carrier screening and prenatal diagnosis, led us in 1975 to set up a preventive programme based on voluntary carrier-screening and prenatal diagnosis[4-5] designed to reduce its incidence. This paper reviews the characteristics and the results of this programme.

Molecular bases of β-thalassemias in the Sardinian population

The first β-globin gene from a Sardinian patient with thalassemia major was described several years ago by sequence analysis, which revealed the presence of a nonsense mutation (CAG→TAG) at codon corresponding to amino acid 39 (codon 39 nonsense mutation).[6] Later, studies carried out by oligonucleotide analysis on either non-amplified or amplified DNA led to the definition of the β-thalassemia mutation in 2,884 chromosomes from subjects of Sardinian descent. The most frequent mutation, accounting for 95.7% of the β-thalassemia chromosomes, turned out to be the codon 39 nonsense mutation, followed by frameshift at codon 6 (codon 6-1bp), which accounted for 2.1% of the β-thalassemia defects.[7] Both mutations produce the phenotype of β°-thalassemia. Homozygous β°-thalassemia most commonly produces the phenotype of thalassemia major. In our population however, in about 8-10% of the cases,[8] homozygous β°-thalassemia results in a disease of moderate severity called thalassemia

* Istituto di Clinica e Biologia dell'Età Evolutiva, Università Degli Studi Cagliari, Italy.

intermedia.[9] Studies carried out in the last few years have shown that the homozygous state for frameshift at codon 6, or the compound heterozygous state for this mutation and codon 39 nonsense mutation, results frequently in a mild phenotype. This may be related to the fact that frameshift at codon 6 is always contained in haplotype IX, which contains a substitution at position -158 5' to the G globin gene, and in conditions of erythropoietic stress seems to be associated with high globin gene output.[10-11] The high globin chain production, by compensating the absent β-chain production, may explain the mild phenotype. In a limited proportion of homozygotes for codon 39 nonsense mutation, the mild phenotype may be explained by the co-inheritance of the deletion of two α-globin structural genes or point mutations in the $\alpha 2$ globin gene. However, in the remaining cases, which represent the large majority, the molecular reason for the mild phenotype has not been elucidated yet.[9] Therefore in our population the prediction of the clinical phenotype on the basis of molecular diagnosis is severely limited.

Atypical β-thalassemia heterozygotes

Heterozygous β-thalassemia commonly results in a hematological phenotype characterized by microcytosis, reduced Hb content per cell, high Hb A_2 and unbalanced globin chain synthesis. Two types of β-thalassemia heterozygotes that do not comply with this definition (atypical β-thalassemia) occur with a relatively high frequency in the Sardinian population. The first, characterized by normal MCV and MCH, and balanced α/β-globin chain synthesis, and defined solely by high Hb A_2 is the result of the co-inheritance of α-thalassemia ($-\alpha/-\alpha$, β-thal βA).[12] Because this type of heterozygous β-thalassemia is easily overlooked in carrier screening by MCH/MCV determination, we carry out Hb A_2 quantitation in each subject. Quantitation of Hb A_2 is obtained automatically by HPLC, which gives highly reliable results.[13] The second type of atypical heterozygous β-thalassemia is characterized by low MCV/MCH, unbalanced globin chain synthesis and borderline to normal Hb A_2, and thus may be confused with α-thalassemia. This phenotype is the result of the double heterozygous state for δ- and β-thalassemia.[14-15] In the last few years, we have defined the δ-thalassemia mutations in a number of these heterozygotes. The most common mutation turned out to be the G$\rightarrow$T substitution at codon $27(\delta^{+27}$ thal.).[16] In a few cases we detected a deletion of 7,201 bp in the β-δ-globin gene region (δ°-thalassemia).[17] The definition of these double δ- and β-thalassemia heterozygotes is, at present, accomplished by dot blot analysis on amplified DNA, with oligonucleotide probes specific for δ^{+27} and deletion δ°-thalassemia.

Carrier screening

Voluntary screening was offered to young unmarried adults, and to prospective parents, primarily to couples during a pregnancy.

Only one member of each couple was tested; the other was tested if the partner's test disclosed a carrier state. The sensitization and involvement of the population included the following: (a) consultation with parents' associations; (b) meetings with community leaders; (c) meetings with physicians, primarily obstetricians and pediatricians, family planning associations, nurses and social workers; (d) education on the inherited anemias in the primary and secondary schools, (e) presentation of the programme to the general public by means of the mass media, and (f) provision of information leaflets at marriage registry offices, general practitioner's offices, and family planning clinics.

Information leaflets provided the following: (a) to whom testing is available, (b) where and how to get testing, (c) heterozygotes are at no disadvantage, (d) description of the homozygous state, and (e) the homozygous state can be prevented safely for the mother.

Before publicizing the programme, we set up adequate facilities to meet the demand for screening as well as for antenatal testing. Any expansion of the educational programme was preceded by a careful calculation of the increased workload and proportional expansion of the laboratory facilities.

Informed consent by the screenee was not requested but, prior to testing, an effort was made to inform each person about the nature of the illness, the meaning of the carrier state, the implication of being a carrier, and the alternatives available to individuals found to be carriers. Once identified, each carrier was informed about the implications of the carrier status for close relatives, and was given simple, clearly written, educational material. Relatives were informed in this way and were given the option to contact the centre if they desired further information or wished to be screened. This strategy helped to multiply the efficacy of screening. By the examination of 167,000 people, we have so far detected 30,500 β-thalassemia carriers (18.3% of those tested) and 1,544 couples at risk. By adding to this figure the number (812) of known couples with children affected by thalassemia major, the total number of couples at risk detected rises to 2,356, which represents 87% of the total predicted on the basis of the carrier rate. The high efficiency of our programme depends on the wide use of inductive screening, namely the extended family examination in all cases in which a person with heterozygous or homozygous β-thalassemia is detected.

According to the results discussed in the previous section on atypical heterozygotes, in all individuals presenting at our genetic clinic, we carry out red-cell indices analysis with an automatic cell counter (Coulter Counter model S plus), and automatic Hb A_2 determination by HPLC and Hb electrophoresis on cellulose acetate. Once the couple at risk is identified, we define the mutation in both parents by dot blot analysis on their amplified leukocyte DNA with ^{32}P or horseradish labelled oligonucleotides probes complementary to either codon 39 nonsense mutation or frameshift at codon

6. Those cases not defined by this approach are tested with oligonucleotides complementary to rarer mutations.

Non-directive counselling is carried out. The counsellor describes the natural history of thalassemia major, explains the available treatment, discusses the various options for preventing the disease, and explains the sampling procedure, its associated risk of fetal loss, and the reliability of the molecular diagnosis.

After the introduction of first-trimester diagnosis 99% of the couples counselled accepted prenatal diagnosis as a means of monitoring for the presence of an affected fetus. In those cases in which a fetus affected by homozygous β-thalassemia was detected, five of 715 (0.7%) women decided to continue the pregnancy because, for ethical reasons, these couples were against the interruption of the pregnancy.

Prenatal diagnosis

From 1977 to 1983 prenatal diagnosis was carried out by globin chain synthesis analysis on fetal blood, and from 1983 to 1988 by oligonucleotide hybridization on electrophoretic-separated DNA fragments.[18] From 1989 we have been using dot blot analysis on amplified DNA with oligonucleotide probes complementary either to codon 39 nonsense mutation or frameshift at codon 6. Fetal DNA is obtained from either amniocytes or chorionic villi. Chorionic villi are sampled by either a transvaginal or a transabdominal approach, which, in our hands, has proved to be a very safe procedure, with a fetal loss of 2%.

Chorionic villi DNA analysis with either amplified or non-amplified DNA has proved to be a very reliable procedure. These has been no misdiagnosis so far. Because of the high efficiency of the polymerase chain reaction used to amplify the DNA, concern has been expressed about the possibility of co-amplification of maternal sequences in such a way as to cause misdiagnosis. The most worrisome occurrence, of course, is co-amplification of maternal sequence in the case of a homozygous fetus, which may thus be mistakenly defined as a heterozygote. We have carefully analysed this possibility by carrying out prenatal diagnosis in duplicate on either non-amplified or amplified DNA, and by splitting the chorionic villi to be amplified into two samples, which are amplified and analysed separately.

In the analysis of 425 cases we have so far detected four in which one amplified sample turned out to be heterozygous β-thalassemia and the other was normal (two cases) or affected (two cases). The fetuses were indeed normal or affected, according to the analysis of non-amplified DNA. These findings clearly indicate that the co-amplification of maternal sequences from maternal leukocytes or decidua may result in misdiagnosis. To avoid this pitfall it could be useful to amplify simultaneously a highly polymorphic VNTR[19], which may show the contribution of two maternal chromosomes and in this way indicate the presence of maternal contamination.

75

Control of β-thalassemia

Carrier screening and prenatal diagnosis have resulted in a rapid decline in the incidence of thalassemia major in the Sardinian population. At present it is 1:1000 live births, with an effective prevention of 90% of the cases predicted on the basis of the carrier rate.

The reasons for the residual cases of homozygous β-thalassemia in the Sardinian population were found to be parental ignorance about thalassemia, in the large majority of cases (67%), followed by false paternity (13%), and a decision not to interrupt pregnancy when prenatal testing has indicated an affected fetus (20%).

Discussion

From our experience in the Sardinian population we may draw several conclusions.

First, carrier screening, counselling and prenatal diagnosis appear to be effective means of controlling an inherited recessive disorder such as β-thalassemia in a population. This experience may be taken as a model for future genetic preventive programmes for other recessive inherited disorders, such as cystic fibrosis. Further improvement may be obtained by educating the population, and especially by introducing the topic of thalassemia into the middle- and high-school curriculum. In the large majority of couples who were ignorant about thalassemia in our population, both parents were very young and had given up attending school very early. Dot blot analysis on amplified DNA with allelic specific probes is a very simple procedure for prenatal diagnosis. It requires a very small amount of DNA, in the order of 5 μg, gives the results within 24 hours of chorionic villi sampling, and may avoid the use of radioactive probes. However, even with the careful dissection of chorionic villi from maternal decidua, especially with very small samples, we have shown in a limited number of cases co-amplification of maternal sequences in such a way as to cause misdiagnosis. To avoid this error, it seems necessary to monitor for the presence of co-amplified maternal sequences by the simultaneous amplification of a polymorphic DNA sequence.

Prenatal diagnosis is nowadays carried out in practically all cases by chorionic villi analysis, because couples who have genetic counselling opt for this procedure instead of amniocentesis, mainly because it can be made at an earlier stage of gestation.[20] For chorionic villi sampling we use the transabdominal approach, which in our experience is associated with less risk of fetal loss than the vaginal route. However, for a final conclusion on this point the results of a randomized study, now in progress, should be awaited.

One of the principles of genetic counselling is the accurate description of the natural history of the disease. Because the clinical phenotype of homozygous β-thalassemia may be heterogeneous and a number of cases, referred to as thalassemia intermedia, may have a mild course not requiring transfusions, it will be useful, in order to improve counselling, to be able to predict the clinical phenotype. However, prediction of the clinical phenotype on the basis of DNA analysis on prospective parents cannot be carried out in our population, because the molecular bases of the mild forms of the β-thalassemia in this population, for the vast majority of the cases, have not yet been defined.[9] In other populations, however, finding very mild β-thalassemia mutations such as the β^{-101} (C→T), β^{-87} (C→G), or the β + IVS-1, nt 6, in both parents, or even in one, may permit the prediction of a mild phenotype in the offspring.[21-24] It is well known that co-inheritance of α-thalassemia with homozygous β-thalassemia may result in a mild clinical phenotype.[25-26] However, this effect of α-thalassemia is not consistent, and thus the investigation of the presence of α-thalassemia in the prospective parents does not seem to be useful for improving genetic counselling.[9]

Future development in this field is related to either the sampling procedure or the method of analysis. The possibility of defining the mutation by the analysis of a limited number of cells, and even from a single cell, may pave the way to earlier diagnosis in pregnancy or pre-implantation diagnosis. It is reasonable to assume that the minimal amount of cells taken by either amniocentesis or chorionic villi at 6-7 weeks gestation may be sufficient to yield enough DNA as a template for DNA amplification[27]. Pre-implantation diagnosis may also be carried out by biopsy at the morula stage after *in vitro* fertilization,[28] or by biopsy of the blastula washed out from the uterine cavity after *in vivo* fertilization.[29]. From the technical point of view, further improvement, especially in rapidity and simplicity, may be achieved by the use of reverse oligonucleotide hybridization, denaturing gradient gel electrophoresis[30-31] or heteroduplex chemical mismatched cleavage analysis.[32-33] Prenatal diagnosis of β-thalassemia has greatly benefited the Sardinian population, both as individuals, by avoiding the tragedy of a son with thalassemia major, and as a society, by avoiding the enormous financial burden created by such a disease. It is to be hoped that in the near future in Sardinia, by further education of the population, thalassemia major may be solely an historical textbook description.

Acknowledgments

We thank Antonietta Sanna, Rita Loi and Sally Harvey for editorial assistance.

This work has been sponsored by the World Health Organization and was supported in part by grants from *Assessorato Igiene e Sanità, Progetto finalizzato Malattie Genetiche di Notevole Rilevanza in Sardegna, CNR contratto n. 89.00307.75,* MPI 40 and 60 per cent 1989, and *CNR Istituto di Ricerca sulle Talassemie e Anemie Mediterranee, Cagliari.*

References

1. Cao, A. et al. Thalassemia types and their incidence in Sardinia. *J. Med. Genet.* 1978; 15:443-7.
2. Thomas, E.D. et al. Marrow transplantation for thalassemia. *Lancet* 1982; ii:227-9.
3. Lucarelli, G. et al. Bone marrow transplantation in patients with thalassemia. *N. Engl. J.Med.* 1990; 322:417-21.
4. Cao, A. et al. Prevention of homozygous β-thalassemia by carrier screening and prenatal diagnosis in Sardinia. *Am. J. Hum. Genet.* 1981; 33:592-605.
5. Cao, A. et al. The prevention of thalassemia in Sardinia. *Clin. Genet.* 1989; 36:277-82.
6. Trecartin R.F. et al. $\beta°$-thalassemia in Sardinia is caused by a nonsense mutation. *J. Clin. Invest.* 1981; 68:1012-7.
7. Rosatelli, M.C. et al. β-thalassemia mutations in Sardinians: implications for prenatal diagnosis. *J. Med. Genet.* 1987; 24:97-100.
8. Galanello, R. et al. Clinical presentation of thalassemia major due to homozygous $\beta°$-thalassemia. *Nouv. Rev. Fr. Hematol.* 1981; 23:101-6.
9. Galanello, R. et al. Molecular analysis of $\beta°$-thalassemia intermedia in Sardinia. *Blood* 1989; 74:823-7.
10. Gilman, J.G. & Huisman T.H.J. DNA sequence variation associated with elevated fetal G globin gene production. *Blood* 1985; 66:783-7.
11. Miller, B.A. et al. Molecular analysis of the high-hemoglobin-F phenotype in Saudi-Arabian sickle cell anemia. *N. Engl. J. Med.* 1987; 316:244-50.
12. Melis, M.A. et al. Phenotypic effect of heterozygous α and $\beta°$-thalassemia interaction. *Blood* 1983; 62:226-9.
13. Mosca, A. et al. An evaluation of the DIAMAT HPCL Analyser for simultaneous determination of hemoglobin A₂ and F. *J. Auton. Chem.* In press.
14. Paglietti, E. et al. Genetic counselling and genetic heterogeneity in the thalassemias. *Clin. Genet.* 1985; 28:1-7.
15. Galanello, R. et al. Interaction between deletion δ-thalassemia and β-thalassemia (codon 39 nonsense mutation) in a Sardinia family. *Proceedings Sixth Conference on Hemoglobin Switching.* A.R. Liss, New York. 1989.
16. Moi, P. et al. Delineation of the molecular basis of δ- and normal HbA₂ β-thalassemia. *Blood* 1988; 72:530-3.
17. Galanello, R. et al. Deletion δ-thalassemia: the 7.2 Kb deletion of Corfu δ-β-thalassemia in a non β-thalassemia chromosome. *Blood*. In press.
18. Cao, A. et al. The prenatal diagnosis of thalassemia. *Br. J. Hematol.* 1986; 63:215-20.
19. Horn, G.T. et al. Amplification of a highly polymorphic VNTR segment by the polymerase chain reaction. *Nucleic Acid Res.* 1989; 17:2140.
20. Cao, A. et al. Chorionic villus sampling and acceptance rate of prenatal diagnosis. *Prenat. Diag.* 1987; 7:531-3.
21. Gonzalez-Redondo, J.M. et al. A C→T substitution at nt — 101 in a conserved DNA sequence of the promoter region of the β-globin gene is associated with "silent" β-thalassemia. *Blood* 1989; 73:1705-11.
22. Ristaldi, M.S. et al. The C→T substitution in the distal CACCC box of the β-globin gene promoter is a common cause of silent β-thalassemia in the Italian population. *Br. J. Haematol.* 1990; 74:480-6.
23. Rosatelli, M.C. et al. Thalassemia intermedia resulting from a mild β-thalassemia mutation. *Blood.* 1989; 73:601-5.
24. Tamagnini, G.P. et al. β+thalassemia Portuguese type: clinical, haematological and molecular studies of a newly defined form of β+thalassemia. *Br. J. Haematol.* 1983; 54:189-200.
25. Weatherall, D.J. et al. Molecular basis for mild forms of homozygous β-thalassemia. *Lancet.* 1981; 1:527-9.
26. Wainscoat, J.S. et al. Globin gene mapping studies in Sardinian patients homozygous for $\beta°$-thalassemia. 1983; *Mol. Med.* 1:1-11.
27. Li, H. et al. Amplification and analysis of DNA sequences in single human sperm and diploid cells. *Nature* 1988; 335:414-7.

28. Handyside, A.H. et al. Biopsy of human preimplantation embryos and sexing by DNA amplification. *Lancet* 1989; 1:347-9.
29. Buster, J.H. & S.A. Carson. Genetic diagnosis of the preimplantation embryo. *Am. J. Med. Genet.* 1989; 34:211-6.
30. Fisher, S.G. & L.S. Lerman. DNA fragments differing by single base-pair substitutions are separated in denaturing gradient gels: correspondence with melting theory. *Proc. Natl. Acad. Sci. USA.* 1983; 80:1579-83.
31. Sheffield, V.C. et al. Attachment of 40-base-pair G + C-rich sequence (GC-clamp) to genomic DNA fragments by the polymerase chain reaction results in improved detection of single-base changes. *Proc. Natl. Acad. Sci. USA.* 1989; 86:232-6.
32. Cotton, R.G.H. et al. Reactivity of cytosine and thymine in single-base-pair mismatches with hydroxylamine and osmium tetroxide and its application to the study of mutations. *Proc. Natl. Acad. Sci. USA.* 1988; 85:4397-401.
33. Grompe, M. et al. Scanning detection of mutations in human ornithine transcarbamoilase by chemical mismatch cleavage. *Proc. Natl. Acad. Sci. USA.* 1989; 86:5888-92.

PRESYMPTOMATIC TESTING FOR HUNTINGTON'S DISEASE: HARBINGER OF THE NEW GENETICS

N. Wexler[*]

There is an allure to trying new technologies. We flex our laboratory muscles and prepare to test our latest diversions. And there is something particularly aesthetic about genetic diagnosis, with those clean, impersonal bars stretching across the lanes at their appointed positions. Without words, they can reveal past couplings or confess extramarital transgressions. And they are clairvoyant, unpeeling shrouds from the future as well as the past.

When a DNA marker was discovered in 1983 closely linked to the Huntington's disease gene, a world was opened up, not only for families suffering from Huntington's disease, but for all those with genetic illness.[1-4] Localizing a gene whose chromosomal origin was unknown confirmed the practical value of a novel strategy applicable to almost all hereditary disorders. Gene mapping, aimed at spanning the genome at regular intervals or the fine point charting of interstices between intervals, is proceeding at a furious pace and a complete map of the human genome will be in place in the not-too-distant future. But as this molecular cartography advances, clinical medicine is turning topsy-turvy in its wake.

Huntington's disease is an autosomal dominant neurodegenerative disease. Appearing usually in the third or fourth decade of life, it can begin in early childhood or old age. A triad of disturbances is pathognomonic: uncontrolled involuntary movements usually including chorea, intellectual decline, and psychiatric disturbance, mostly depression. The disease progresses for 20 years toward an inevitably fatal outcome and therapy is palliative at best.[5]

Families, scientists and clinicians involved with research on, or treatment of, Huntington's disease have been aware that their development of a counseling and diagnostic program using DNA markers could be as precedent-setting as the discovery of the marker itself. They have been trying to act with caution and sensitivity while at the same time making the test available.

The tests offered

Presymptomatic and prenatal testing can now be carried out with a number of different markers to enable risks to be changed from 50% to 96% or higher of being positive or negative for the Huntington's disease gene.[6-10] Even though there has been some uncertainty as to the exact location of the Huntington's disease gene within 4p16.3, the most telomeric band on chromosome 4, there are sufficient markers to alter a person's risk

---

[*] Professor, Clinical Neuropsychology, College of Physicians and Surgeons of Columbia University, New York, U.S.A.

significantly if a family is genetically informative.[11,12] Initially, less than half of all families listed on the National Huntington's Disease Roster at Indiana University had such an appropriate structure.[13] Owing to the increased informativeness of recently developed markers, recent estimates suggest that up to 75% of individuals coming for testing will have genetically informative families.[14]

Prenatal diagnosis

Two types of prenatal diagnostic tests are available:
1) a nondisclosing prenatal or "exclusion test," and
2) a fully disclosing diagnostic test.[15-17] In a nondisclosing prenatal test, the risk status of the at-risk parent is not altered, only the risk of the fetus. DNA from the fetus is checked for the presence of the grandparental chromosome from the affected or from the unaffected grandparent. In the first instance, the fetus would have a 50% risk, or the same risk as the at-risk parent who carries the same chromosome. In the second instance, the fetus would have only a negligible risk (approximately 2%, due to the possibility of recombination). The option of a nondisclosing prenatal test can be offered to those at risk who do not wish to know their genetic status or who do not have sufficient family members to make a full disclosing test possible. A nondisclosing prenatal test can be carried out with DNA from only the expectant parents, the fetus, and a single parent, either healthy or affected with the disease, of the at-risk individual.

If a family is genetically informative, and the couple wishes to know the information, the fetus can be tested, with full disclosure of its genotype. Some parents may choose to proceed with full disclosure after a previous nondisclosing test has revealed a fetus with a 50% risk. If the fetus proves to have a negligible risk the genetic identity of the at-risk parent is still protected. If the fetus is found most likely to carry the HD allele, parent and child are diagnosed simultaneously in a double tragedy.

One problem which arises during prenatal testing is that the analysis of samples is time-consuming. Couples who embark on full testing of a fetus, even after chorion villus sampling at 8-12 weeks, may endure very late and often psychologically traumatic terminations.

Presymptomatic testing

Presymptomatic and prenatal testing is now being offered in 22 centers in the United States, as well as several centers in the United Kingdom and Canada.[18-25] Canada has formed the Canadian Collaborative Study of Predictive Testing for Huntington's Disease and has created 14 testing centers across Canada. France is just embarking on a program in Paris, and other countries, such as Germany and Scandinavia, may be beginning. I have directed a small pilot project providing testing at the College of Physicians and Surgeons of Columbia University since 1987 under the auspices of the Robert Wood Johnson Foundation. The following observations reflect our experiences, together with those from other centers.

81

Guidelines for testing

There are variations among countries and centers but the following represents some key elements of the basic protocol which is followed, at least in the United States and Canada:

1) There must be a minimum of three to six separate counseling sessions before diagnostic information is delivered. Each session should be several hours long. Intensive counseling regarding motivations and preparation for testing is the most essential element of the entire protocol. Post-test counseling *must* be part of any protocol but clients sometimes prefer to seek counseling closer to home.

2) Potential clients should be evaluated neurologically, neuro-psychologically and psychiatrically.

3) Relatives at risk or symptomatic who are donating a DNA sample for a client to be tested should be evaluated neurologically as well. A diagnosis should not be accepted on the basis of hearsay evidence, even from a family member. Corroboration should be sought from the diagnosing physician and a re-evaluation should be arranged if there is any doubt. If persons at risk cannot or will not be examined, their risk should be assigned very conservatively in the linkage analysis. The disease should be confirmed in at least one relative by autopsy diagnosis or by very reliable neurological examination.

4) Clients found to have significant psychiatric disorders, particularly a history of suicide or severe depression, or those undergoing stressful life circumstances causing emotional upheaval, such as divorce or a death in the family, are not suitable testing candidates.

5) Diagnostic information must always be given in a face-to-face session, *never* over the telephone. Even if the outcome is genetically uninformative, clients need an opportunity to discuss what this information may mean.

6) Most programs require or strongly urge that clients be accompanied by a companion to at least one counseling session and at the disclosure session.

7) Long-term follow-up is essential, particularly for those who test positive for the gene. As the time draws near when symptoms are likely to appear, clients need to know that they have a relationship with a supportive therapeutic individual or group.

8) Some programs require that a client contact a psychotherapist prior to receiving diagnostic information. Other programs provide the therapists. These therapists can continue to see clients, particularly after a positive diagnosis.

9) All DNA determinations must be carried out independently at least twice. If contradictions appear, new DNA samples are collected. Some centers collect two independent blood samples. If blood has been donated from relatives for research purposes these samples must be re-collected unless explicit permission is given for them to be used in diagnostic testing. Even then it is best to collect new samples on crucial individuals.

10) A genetic linkage computer analysis of haplotypes generated *must always* be conducted. Diagnostic information must *never* be given on the basis of a visual analysis of the gels alone.

11) If siblings of a client need to be analyzed to determine phase or to reconstruct the haplotype of a deceased parent, the identities of these siblings should be confidential and data analyzed anonymously. This prevents those providing counseling from inadvertently receiving unwanted and inappropriate information.

12) Testing should be available only to persons aged 18 years or older who can give informed consent. One potential complication of this guideline may occur if parents have a nondisclosing prenatal test and choose to maintain a pregnancy in which the fetus is found to have a 50% risk. If the at-risk parent develops Huntington's disease, the child is *de facto* diagnosed. Another ethical quandary may occur when couples who wish to adopt a child at risk insist on testing as a condition of adoption.

13) All testing must be *totally voluntary* and the results remain *totally confidential, even to other family members.*

Attitudinal surveys prior to test availability

Following the announcement of the discovery of the marker, attitudinal surveys were conducted of families with Huntington's disease. Some surveys were aimed only at those at risk while others solicited opinions from the entire family. There was a wide range of outcomes, from as low as 40% of the at-risk population interested in testing to as high as 100%.[26-32] A few investigators found that some people, once positive about testing, had changed their minds with the pending availability of the test. A Dutch survey concluded that those more knowledgeable about the test itself were less eager to be tested.[33] In all studies, level of education or income, occupation or marital status made no difference.

Who comes for testing?

There are four outcomes from presymptomatic linkage testing for the Huntington's disease allele. A person can have a high probability of being positive or negative for carrying the gene. The test can also prove genetically uninformative. Occasionally, nonpaternity is revealed, which indicates that the testee is not actually at risk.

Every at-risk person who seeks testing has a unique story but there are some generalizations one can make about characteristics and motivations. Individuals requesting testing can be grouped into four general categories: young adults, older at-risk parents, the offspring of people newly diagnosed, and clinically symptomatic individuals who do not recognize their affected status.

Young adults

One group is comprised of young adults in their twenties and thirties who are planning, or have recently entered into, new careers, or to be married and have children. It was initially thought that this group would be the largest, but experience has yet to determine whether this will be the case. These people are perhaps most affected by whether or not they are gene carriers, as the test will greatly influence their decision-making. Many are enthusiastic, resilient and full of youthful energy. This group also has the highest genetic risk as most of them are young and still have a 50% probability (which declines with age) of having inherited the disease. They have the most to gain and the most to lose. Many of them are also filled with the adolescent's sense of invulnerability — "It can't happen to me." They feel so healthy, well coordinated, functional and perfectly fine that the prospect of having the illness is more or less an intellectual proposition. At the same time, they scan themselves incessantly for initial signs. A sense of invulnerability and adolescent bravado may push them to take the test without adequate psychological preparation for the possibility that the gene is present. They are often full of plans for what they will do if the gene is not there, but are vague and unsettled about a life waiting for the disease to make its appearance. News of a positive result can be shocking and traumatic even with adequate psychological rehearsal and preparation.

Young adults have the longest to wait in the "presymptomatic state", before the disease becomes manifest. For most, this seems to be a boon — the disease is a long way off. But for others, these may be years of anxiety, dread and ambiguity, of hypochondriacal concern and hypervigilance for symptoms. Because of the newness of the testing programs, very few individuals have had to live even three years with the knowledge of a positive test outcome. Some may deny the appearance of symptoms when they do begin and others may shut down early on a healthy life, ending the ambiguity and the waiting by becoming a patient prematurely.

For some young people, the knowledge that they are destined to develop Huntington's disease arrives before a career is chosen or a marriage bond sealed. Gene carriers may pull back on wishes to become a physician, an astronaut, a Wall Street financier, any profession requiring a great deal of training, coordination and judgment. And some would argue that this is all for the best in that they will not pose a hazard for themselves and others. Yet many with Huntington's disease have enjoyed great success and satisfaction in a variety of careers prior to becoming incapacitated. Some presymptomatic carriers may constrict their lives unnecessarily, while others may be galvanized into leading a fuller life, knowing that time is limited.

Marital planning is an important impetus for testing. Often at-risk individuals are more eager to clarify their risk status than are the prospective spouses. Persons at risk want their prospective spouses to know what is in store for them. But some prospective spouses explicitly do not want to know. They feel bound to honor the engagement even more if the outcome is a probable gene-positive test, so they prefer to get married in hope. Many at risk feel guilty about imposing their uncertain and possibly difficult future on their spouses. A positive diagnosis may make single presymptomatic individuals retreat even further from marriage and intimacy.

The timing of Huntington's disease, with its late onset and prolonged course, can cause problems in adjusting to diagnostic testing. Many young adults who are eager for testing have parents who have just died or are in the terminal phases of the illness. It can be traumatic to nurse a dying parent after learning that this is to be one's own fate. At-risk individuals are frequently the primary carers of their parents with Huntington's disease. Knowing that they have inherited the same gene may make this task doubly painful and cause role readjustments for all family members.

Older at-risk parents: the altruistic testee

The second group interested in presymptomatic testing consists of at-risk individuals in their fourth through sixth decades who are parents of children approaching young adulthood. These older people at risk often do not want predictive information for themselves. Left to their own devices, they would not get tested. They have led full lives, made decisions as best they could, and gained a certain feeling of confidence as the disease has not yet appeared. They know, however, that if they are tested and found free of the gene their children are relieved of any risk.

Many parents are willing to subject themselves to the testing as a gift to their children. Given their age, their genetic risk is already reduced, which augurs well for a good outcome. However, to a person who has achieved some sense of freedom from the threat of Huntington's disease, the information that the gene is most probably present may be all the more devastating. Also, the disease would be more likely to appear soon, depending on their age. On the other hand, the fact that they have lived without symptoms for many years and accomplished much may bring some solace and diminish the impact of bad news. We must learn whether taking the test for altruistic motivations produces different needs and stresses, both pre-and post-testing.

The newly at-risk

The third category is comprised of individuals who have just learned that they are at risk, regardless of their age. A fairly universal reaction of people learning for the first time of their genetic risk is to want to resolve their doubt immediately. This novel situation of risk is so ambiguous and so uncomfortable that it is intolerable. This group is perhaps the most vulnerable, for they have not yet had time to realize the implications of the disease and the test. Many have never seen the illness in its terminal stages. It can be disastrous if they get this preview of their own end after they find out that they themselves are positive for the gene. It takes many years to adjust to being at risk; most of the newly at risk are still reeling from this shock. They must be helped to adjust to the risk situation and learn more about the disease before proceeding to presymptomatic testing.

The clinically affected

A small but significant number of people coming for testing are already clinically affected. Some may have an inkling that they are affected and they come for confirmation; others are surprised. Some genuinely do not want to know that they are affected, only whether they will be affected in the future; it is best to recommend that they postpone testing.

The family as a testing partner

Testing does not occur in a vacuum. Samples must be collected from critical relatives, whose permission is required for testing, and each family member has his or her own stake in the information delivered.

Parents must provide a sample for the test to be successful. Most times the healthy parent has spent years caring for an ill spouse and may be looking forward to a respite before children begin the same fatal trajectory. It is also horrifying for them to face the prospect that they may die and be unable to care for sick children. Parents affected by Huntington's disease may feel especially guilty, knowing that they have transmitted the fatal gene to their children.

An ethical and psychological dilemma has arisen in several testing centers (Myers Brandt, personal communication). A parent has consented to give a sample for one offspring but not for another, feeling that the children differed in their ability to handle diagnostic information. Once the parent's sample is used in one test and the genetic haplotype determined, it is not needed for any other offspring's test. So despite the parent's wishes, technically the test could be done for all interested offspring. Clearly an ethical problem exists: should testing continue contrary to the parent's wishes or stop, thereby depriving all offspring of information about genotypes. The question is raised of who "owns" a parent's genetic haplotype? Whose wishes have precedence, the at-risk individual wishing personal testing or the parent, who may be acting out of malignity or compelling concern? The best counseling alternative is to endeavor to bring family members to a resolution satisfactory to everyone before testing any of the offspring.

Testing one sibling has repercussions for all. Bringing the test close to home usually forces siblings to decide more actively whether they wish to be tested. Sometimes their genetic information may be required for their sibling's test, to reconstruct the haplotype of a deceased parent, for example. The optimal timing for one sibling to be tested may conflict with another's. Deciding when to take the test can be almost as crucial as whether to take it.

Family secrets

Siblings may entertain the fantasy that one of them could be tested and the others in the family would not know. In some families, this may be the case. However, most family members in reasonable contact with each other find such momentous news hard to hide. Changes in plans — for example, a decision to have children or a prolonged absence or period of depression — are revealing signs. And keeping the news from family members may deprive someone newly presymptomatic of necessary support.

If siblings know the outcome of one sibling's test it can affect them, whether or not they choose to be tested. Even though "chance has no memory," siblings often feel that if one goes free their risk increases, or, conversely, if one is diagnosed their risk diminishes. Although each sibling has an independent risk, siblings often feel that their fates are intertwined. Sometimes all siblings in a family are diagnosed positively or negatively, which can be catastrophic or joyous for all. Disparities in test outcome can exacerbate sibling rivalries and family tensions as well as draw families closer together.

Motivations for diagnostic testing

At-risk individuals can come for testing because they want to know either outcome: Huntington's disease gene positive or negative. Others want to know whether they will be free of the disease and are willing to risk hearing bad news for the joy of hearing good.

People come with a variety of motivations, some concerned with planning and others with ending ambiguity. It remains to be seen how much their stated intentions and feelings prior to testing accord over time with their actions following receipt of diagnostic information.

Do people really want to know the truth? A woman came to our testing clinic, highly functional, intelligent, a working wife and mother, and lamented that her family was not genetically informative enough for her to be tested. She stated that she would definitely like to be tested as this would help her advise her family better. This woman had been symptomatic for about five years but never diagnosed. Her statements about why she wanted the test were cogent and persuasive. She wanted to know the truth. Yet something in her very powerfully did not want to know the facts. The truth was that she functioned better in denial; when the denial was finally cracked by a clinical diagnosis, she and her family were devastated.

Testing programs are too recent to have much experience with long-term follow-up. Certain general trends seem to be appearing in the lives of those undergoing testing, which vary according to the diagnostic outcome.

The "escapee"

This group has been the largest across all testing centers. They also seem to undertake the most active changes in their lives. Some are freed to have children or marry or get divorced. Some change jobs or go back to school. It appears that many people at risk unduly restrict themselves from options which they then feel free to pursue once the specter of the disease is lifted.

Many who test most likely negatively for the gene also forget that some risk remains. Although those who test most likely gene-positive cling to

87

the hope that a genetic recombination is responsible for producing their untoward outcome, "escapees" "forget" that the test is not 100% reliable.

"Escapees" with siblings or other relatives are acutely aware that others in their family may not prove so lucky. Some have expressed classic "survivors' guilt," vowing to devote themselves to caring for relatives should they fall ill.

Those most likely gene-positive

This group is the smallest and is being watched the most carefully. There has been one reported instance of attempted suicide by a person who came for testing and discovered she was symptomatic.[34] Another person had to be briefly hospitalized for depression after a test outcome with a high probability of being positive.

Hayden has recently reported on longitudinal follow-up of Beck Depression scores of both Huntington's disease patients and persons mostly likely gene-positive. The gene-positive group had fairly stable Beck Depression scores, while the Huntington's disease patients had quite high scores.[35] This should warn us that those with a positive test may be holding depression at bay by thinking that the disease is far in the future, but that as they grow older their risk for depression will increase.

Non-informative test

This group seems to blossom rather like those who receive good news. They get on with their lives and adopt children or change jobs and generally behave as if there has been some resolution of their risk status. It is as if they tried to do all they could to contend with their question of risk and now they can focus on living.

What is an informative test?

Testing centers vary in the information they give to testees. Some centers, such as those at Johns Hopkins, Harvard and Columbia Universities, will consider a test genetically informative if it can change a client's risk to 95% or higher in either direction. Below this level of confidence, the test is too vulnerable to such factors as variation in the ages of onset in the family and other data which tend to be murky. The Canadian Collaborative Study group will provide information on any change in risk levels (M. Hayden, personal communication). My experience with clients in our testing program is that they interpret any change as if it were liberating or ominous, and tend to treat all risk alterations as if they were 100% certainties.

Huntington's disease and AIDS

Analogies can be drawn between presymptomatic testing for Huntington's disease and testing for HIV positivity. In each instance the information to be conveyed is potentially devastating. We still do not know what portion of people who have an HIV positive test eventually develop AIDS. While a diagnosis of Huntington's disease gene positivity carries with it certain lethality, the disease is often further into the future. Both diagnoses bring uncertainty and fear.

Until recently, the Gay Men's Health Coalition was not actively encouraging HIV testing as long as people took precautions. However, with the advent of zidovudine, there is now an appropriate treatment to begin as soon as possible after HIV infection. Attitudes toward testing have changed. Perhaps this will be the case when a treatment is available for Huntington's disease as well.

Genetic illnesses are also sexually transmitted diseases in the sense that only through intercourse are they passed on. As legislation or policies are developed that pertain to HIV infection, those with, or presymptomatic for, genetic diseases may be enmeshed, for better or worse.

Society and the test

I have been focusing mainly on the individual and family aspects of presymptomatic testing for Huntington's disease. There are also important societal implications. For people to decide whether they want to be tested, they should know the legal, financial and social repercussions. Many of these are unexplored areas. How coercive might employers become, knowing the at-risk status of an employee, in pushing that employee to be tested? Can tests be required as a condition of hiring? Will an employer assess that certain positions — neurosurgeon, pilot, construction worker — are too hazardous for someone positive for the gene or even at risk, even though their performance is unchanged? What legal recourse and protections do people have who are presymptomatic? The Americans With Disabilities Act, recently passed by the United States Congress, should provide some protection against discrimination by employers.

Insurance companies have not yet decided what attitude to take to confront the rise in the availability of genetic tests. It is not clear whether they will reimburse for such tests, whether they will require knowing the diagnostic outcome, or whether those found to be positive, for Huntington's disease, for example, will lose their insurance or be required to pay exorbitant premiums. Will those found to be free have a lowering of their premiums? As it is now, many people who are at risk for Huntington's disease and not part of a group health policy are uninsurable.

When the gene is known

Presymptomatic testing for Huntington's disease at present takes from several months to over a year. DNA samples must be collected from relatives, who must be neurologically assessed, the counseling and evaluations are extensive, and time passes. The necessary counseling is often difficult and raises complex psychological, ethical, legal and moral conundrums.[36-37] This drawn-out process acts as a brake on precipitous action and allows a client to live for some time experimenting with both positive and negative outcomes. Couples and families have time to think and discuss with each other. Sometimes a relative whose blood sample is being sought will dissuade an at-risk person from being tested. It takes time to delve through all the layers of defenses that people at risk have built over time to enable them to cope and allow them to feel the emotional reality of a negative or positive test result.[38]

Once the gene is identified, testing will be much more rapid and accurate and less expensive. We do not yet know how many different mutations of the Huntington's disease gene there may be, even though there is no evidence of locus heterogeneity.[39] It may still be necessary to obtain some samples from relatives to determine just which mutation is present in a family, but the very time-consuming linkage typing of relatives will be obviated.

Will this rapidity be a boon? Just because the testing process is faster and cheaper, the information for clients is no less long-lasting or momentous. Failure to insist that people take the time to digest the possibility of hearing potentially catastrophic news could result in many more casualties of testing than we have yet seen.

Genes and mores: givens and chosens

As the titer of HIV rises in our international bloodstream and as more genetic tests are being developed, our sense of endangerment, embattlement and self-righteous defensiveness may also rise. We face a critical time in educating the public and in ensuring that good and culturally appropriate genetic services are available to all people without regard to class or race.

A young man from Germany, at risk for Huntington's disease, commented on presymptomatic testing: "We must be exceedingly careful with any system of records. We must remember that World War II was not so long ago — and that Fascism is with us today." We cannot permit concentration camps, literal or figurative, for the ill or genetically stigmatized. We are embarking upon an experiment in improving health in the midst of a world of fragile peace, a world suffocating with sufficient nuclear weaponry to rob any disease of its target. Technological developments and the Human Genome Project are providing us with unparalleled opportunities to understand hereditary disease, but we must remember the camps and never pretend not to see.

Acknowledgements

I am grateful for the collaboration and expertise of Graciela Penchaszadeh and to Judith Lorimer and Edith Shackell for editorial assistance. This work was supported by a grant from the Robert Wood Johnson Foundation.

References

1. Gusella JF, NS Wexler, PM Conneally, et al. A polymorphic DNA marker genetically linked to Huntington's disease. *Nature* 1983; 306:234-38.
2. Wexler NS, PM Conneally, D Housman and JF Gusella. A DNA polymorphism for Huntington's disease marks the future. *Arch. Neur.* 1985; 42:20-4.
3. Wexler NS. Genetic jeopardy and the new clairvoyance. *Prog. Med Genet.* Vol. 6. A. Bearn, B. Childs and A. Motulsky (eds.). New York, Praeger Press, 1985.
4. Wexler NS. The Oracle of DNA. *Molecular Genetics in Diseases of Brain, Nerve, & Muscle.* L.P. Rowland, D.S. Wood, E.S. Schon, S. DiMauro, (eds.), Oxford University Press, New York, 1989.
5. Wexler NS. Huntington's disease. *Current Opinion in Neurology and Neurosurgery.* S. Fahn and J. Jankovic (eds.). London, Gower Academic Journals, 1:3, 319-23, 1988.
6. Conneally PM, MR Wallace, JF Gusella, NS Wexler. Huntington's disease: estimation of heterozygote status using linked genetic markers. *J. Genet. Epidem.* 1984; 1:81-8.
7. Conneally PM, JF Gusella, NS Wexler. Huntington's disease: linkage with G8 on chromosome 4 and its consequences. *Prog. Clin. Biolog. Research* 1985; 77:53-60.
8. Conneally PM, JF Gusella, NS Wexler, Huntington's disease: genetics, presymptomatic and prenatal diagnosis. *Nucleic Acid Probes and Diagnosis of Human Genetic Diseases.* A.M. Willey, Ed., New York, Alan Liss, Inc., 137-52, 1988.
9. Gilliam TC, RE Tanzi, JL Haines, et al. Localization of the Huntington's disease gene to a small segment of chromosome 4 flanked by D4S10 and the telomere. *Cell* 1987 Aug 14;50 (4): 565-71.
10. MacDonald ME, SV Cheng, M Zimmer, et al. Clustering of multi-allele DNA markers near the Huntington's disease gene. *J. Clinical Investigation* (in press).
11. MacDonald ME, JL Haines, J Zimmer, et al. Recombination events suggest potential sites for the Huntington's disease gene. *Neuron* 1989; 3:183-90.
12. Roberts L. Huntington's gene: so near, yet so far. *Science* 1990; 247:624-7.
13. Farrer LA, RH Myers, La Cupples, PM Conneally. Considerations in using linkage analysis in a presymptomatic test for Huntington's disease. *J. Med. Genet.* 1988; 25:577-88.
14. Brock DJH, ME Mennie, A. Curtis, et al. Predictive testing for Huntington's disease with linked DNA markers. *Lancet* 1989; 2:463-66.
15. Fahy M, C Robbins, M Bloch, et al. Different options for prenatal testing for Huntington's disease using DNA probes. *J. Med Genet.* 1989; 26:353-357.
16. Quarrell OWJ, AL Meredith, A Tyler, et al. Exclusion testing for Huntington's disease in pregnancy with a closely linked DNA marker. *Lancet* 1987; 1:1281-3.
17. Millan FA, A Curtis, M Mennie, et al. Prenatal exclusion testing for Huntington's disease: a problem of too much information. *J. Med. Genet.* 1989; 26:83-5.
18. Meissen GJ, RH Myers, CA Mastromauro, et al. Predictive testing for Huntington's disease with use of a linked DNA marker. *N. Engl. J. Med.* 1988; 318:535-42.
19. Brandt J, KA Quaid, SE Folstein, et al. Presymptomatic diagnosis of delayed-onset disease with linked DNA markers. *JAMA* 1989; 261:3108-14.
20. Quaid KA, J Brandt, RR Faden, SE Folstein. Knowledge, attitude and the decision to be tested for Huntington's disease. *Clin. Genet.* 1989; 36:431-438.
21. Morris MJ, A Tyler, L Lazarou, et al. Problems in genetic prediction for Huntington's disease. *Lancet* 1989; 2:601-3.
22. Craufurd D, A Dodge, L Kerzin-Storrar, et al. Uptake of presymptomatic predictive testing for Huntington's disease. *Lancet* 1989; 2:603-5.
23. Curtis A, F Millan, S Holloway, et al. Presymptomatic testing for Huntington's disease. *Hum. Genet.* 1989; 81:188-90.
24. Mennie ME, SM Holloway, DJH Brock. Attitudes of general practitioners to presymptomatic testing for Huntington's disease. *J. Med. Genet.* 1990; 27:224-7.
25. Stern R, and R Eldridge. Attitudes of patients and their relatives to Huntington's disease. *J. Med. Genet.* 1975; 12:217-33.
26. Teltcher B, and S Polgar. Objective knowledge about Huntington's disease and attitudes toward predictive testing of persons at risk. *J. Med. Genet.* 1981; 18:31-9.
27. Schoenfeld M, RH Myers, B Berkman, E Clark. Potential impact of a predictive test on the gene frequency of Huntington's disease. *Am. J. Hum. Genet.* 1984; 18:423-9.

28. Kessler S, T Field, L Worth, H Mosbarger. Attitudes of persons at risk for Huntington's disease toward predictive testing. *Am J. Med. Genet.* 1987; 26:259-70.
29. Markel DS, AB Young, JB Penney. At risk persons' attitudes toward presymptomatic and prenatal testing of Huntington's disease in Michigan. *Am J. Med. Genet.* 1987 Feb; 26 (2):295-305.
30. Mastromauro C, RH Myers, B Berkman. Attitudes toward presymptomatic testing in Huntington's disease. *Am. J. Med. Genet.* 1987; 26:271-82.
31. Meissen GJ, Rl Berchek. Intended use of predictive testing by those at risk for Huntington's disease. *Am. J. Med. Genet.* 1987; 26:283-93.
32. Kessler S. Psychiatric implications of presymptomatic testing for Huntington's disease. *Amer. J. Orthopsychiat.* 1987; 57 (2):212-9.
33. Evers-Kiebooms G, A Swerts, JJ Cassiman, et al. The motivation of at-risk individuals and their partners in deciding for or against predictive testing for Huntington's disease. *Clin. Genet.* 1989; 35:29-40.
34. Hayden M. *Am. Soc. Hum. Genet.* 1988.
35. Hayden M. *First Int'l Cong. Mvt. Dis.* Apr. 25-27, 1990.
36. Chapman MA. Invited editorial: Predictive testing for adult-onset genetic disease: ethical and legal implications of the use of linkage analysis for Huntington's disease. *Am. J. Hum. Genet.* 1990; 47:1-3.
37. Huggins M, M Bloch, S Kanani, et al. Ethical and legal dilemmas arising during predictive testing for adult-onset disease: the experience of Huntington's disease. *Am. J. Hum. Genet.* 1990; 47:4-12.
38. Wexler NS. Genetic 'Russian Roulette': The experience of being at risk for Huntington's disease. *Genet. Counsel.: Psychological Dimensions.* S. Kessler (Ed.). New York, Academic Press, 1979.
39. Conneally PM, JL Haines, RE Tanzi, NS Wexler, et al. Huntington's disease: no evidence for locus heterogeneity. *Genomics* 5, 1989.

GENETIC ANALYSIS OF FAMILIAL AMYLOIDOTIC POLYNEUROPATHY

Y. Sakaki[*]

This paper summarizes the results of a genetic analysis of familial amyloidotic polyneuropathy (FAP) carried out in Japan and, in collaboration with a Portuguese group, in Europe. The data show that FAP can be diagnosed with very high reliability even at the presymptomatic and prenatal stages. However, the clinical application of the genetic diagnosis is very difficult, as FAP is an adult-onset and fatal disease. Problems in the genetic testing for FAP will be discussed.

FAP and transthyretin

Familial amyloidotic polyneuropathy is a fatal, adult-onset disease showing autosomal dominant inheritance, and characterized clinically by the systemic deposition of amyloid fibrils and progressive disorder of the peripheral nerves. It has been found in various parts of the world, including Portugal, Japan, Sweden, Greece, Italy, Brazil and the United Sates of America. The largest focus has been found in Portugal and there are relatively large foci in Sweden and Japan. In Japan, at least 13 apparently unrelated families have been identified and the number of patients is estimated to be several hundred. The onset of the disease typically occurs between the ages of 20 and 40 years, and death follows in about ten years.

Chemical analysis of amyloid and plasma proteins of FAP patients has revealed that the major component of FAP amyloid is a variant type of transthyretin (TTR) with a single amino-acid substitution. To date, 15 distinct variants related to FAP have been reported, and the mutations responsible for those variants have been identified by the cloning and sequencing of the human TTR gene.

Detection of the mutations

Most of the genetic changes that cause the amino-acid substitutions related to FAP lead to the formation of new restriction sites. In the early stages the new restriction sites were detected by the Southern blot analysis. Recently, the PCR (Polymerase Chain Reaction) amplification of appropriate regions of the TTR gene, and subsequent digestion with restriction enzymes, enabled us to detect the mutations by agarose gel electrophoresis without using radioactive probes. The mutation in the amplified samples can also be detected with ^{32}P-labelled allele-specific oligonucleotide (ASO) probes. The PCR amplification procedure allowed us to detect the Val→Met mutation

---

[*] Research Labaratory for Genetic Information, Kyushu University, Fu Kuoka, Japan.

in <0.1 ng of genomic DNA. As described later, this method may be used for the prenatal diagnosis of FAP.

Genetic test for FAP

Since the first genetic test for FAP, in 1984, by Sasaki *et al.*, we have examined 36 FAP families (13 unrelated groups) in Japan by recombinant DNA technology, and all the subjects clinically diagnosed as FAP were shown to have the mutations heterozygously. DNA testing of FAP patients by other workers in Japan, Europe and the United Sates also has shown that all were heterozygous for the mutations except for one homozygous case in Sweden. Thus, the mutations in the TTR gene are very tightly linked to FAP, showing that the genetic defect in the TTR gene is the primary cause of FAP and that there is no genetic heterogeneity. In collaboration with Dr R. Sparkes (UCLA, USA) we mapped the TTR gene to chromosome 18 q11.2-12.1 by *in situ* hybridization. No disease except FAP has yet been genetically mapped in this region.

Prenatal diagnosis of FAP

It is now possible to diagnose FAP by DNA test, even at presymptomatic and prenatal stages, with very high reliability. A Portuguese group and I have recently performed prenatal diagnosis for FAP in two at-risk fetuses by amniocyte DNA analysis. The amniocytes in one of the cases derived from a pregnant asymptomatic woman carrier of the TTR Met 30 mutation. DNA was extracted from amniocytes, and the second exon of the TTR gene was amplified and digested with *Nsi* I. The PCR-amplified DNA (150 bp) from an FAP individual gives origin to two extra fragments, one of 110 bp and one of 40 bp, but the amniocyte DNA did not present these extra bands. This result was later confirmed by the DNA analysis of the newborn. We also analysed this same amplified DNA by hybridization with ASO probes and obtained the same result. In the other case DNA from a sample of amniotic fluid collected at the 14th week of pregnancy was analysed; in this case the father was the carrier of the mutation. The PCR-amplified DNA from amniotic fluid cells (AF) was shown to hybridize with the "normal" and the "FAP" ASO probes. Therefore the fetus should have a normal allele and an FAP allele, and was thus at risk for the disease.

Late-onset cases

The age of onset of Japanese and Portuguese type I FAP is usually during the third or fourth decade, but extensive screening of FAP families has revealed the presence of late-onset cases that had not manifested the symptoms by the age of 60 or even 70 years. No significant difference was observed between normal- and late-onset cases in biochemical properties and serum concentrations of TTR. Late-onset cases often appeared in the same families. In has often been found that the clinical manifestation of FAP is not identical in members of the same family with the same mutation. These findings suggest that, although the primary cause of FAP has become clear, unknown genetic factors other than TTR, or non-genetic factors, also play an important role in the progression of FAP.

Frequency of occurrence of type I FAP

Among the amino-acid substitutions related to FAP, the ^{30}Val→Met has been most commonly found in FAP individuals of various ethnic gorups. The wide geographic distribution of the identical Val→Met mutation has suggested some special significance of the mutation in FAP. To gain an understanding of the origin and epidemiology of the Val→Met mutation, we analysed DNA polymorphisms associated with the TTR gene in several Japanese FAP families and also, in collaboration with Portuguese workers, in FAP families in Europe. Three distinct haplotypes associated with the Val→Met mutation were identified in the Japanese and European FAP families. The results of our analysis of these haplotypes, together with the fact that the G residue at the first letter of the ^{30}Val codon is involved in a CpG mutation hot spot sequence, led us to consider that the Val→Met mutation has occurred frequently in the human population and that type I FAP is not restricted to some specific ethnic groups but will be shown by extensive genetic testing to occur in many races. This is supported by a recent discovery of a Turkish family with FAP.

Practice and problems in genetic testing for FAP

As described above, a genetic test for FAP has been technically established but the clinical application of the genetic diagnosis is very difficult, much as in Huntington's disease. In contrast to cases of such early-onset, severe diseases as Duchenne muscular dystrophy and thalassemia, the individual carrying the mutation can usually live normally for at least 20-30 years and, in some cases, more than 60 years. Young family members learn the clinical course of FAP by observing their parents and relatives, and face the continuous fear that they themselves may have the genetic defect. Genetic testing may release the non-carrier members from this fear but must cause very serious hurt to the carriers. Some geneticists are considering doing the genetic test only for individuals of more than 20-25 years and after extensive genetic counselling. This seems to be a reasonable solution to the problem. For example, many family members of marriageable age visited hospitals for genetic testing in Portugal (more than 2500), and also in Indiana. It should be noted that the patients' association in each country is playing a key role in establishing a system for genetic testing.

Some people argue that FAP can be eliminated from the human population within 20-30 years by the extensive use of prenatal diagnosis. However, the number of FAP patients is not sufficiently high for FAP to be considered a serious social problem. Thus, society may not have a rationale to compel or even offer the systematic use of prenatal diagnosis of FAP.

Only five cases of prenatal genetic testing for FAP have been reported; in all cases the tests were done to prove the efficacy of the method. If an effective therapy were to be developed within the next 20 years, a carrier born now could benefit from it.

Acknowledgements

I thank very much my colleagues in my laboratory for participating in the study on FAP. The data on the genetic screening of FAP in Japan were obtained in collaboration with Drs. H. Matsuo and M. Nakazato (Miyazaki Medical School), I. Goto (Kyushu Univ.), T. Isobe (Kobe Univ.), S. Kito (Hiroshima Univ.), S. Ikeda and N. Yanagisawa (Shinshu Univ.), K Sahashi (Aichi Medical School) and many other physicians in the hospitals. The analysis of European FAP families and prenatal diagnosis were carried out in collaboration with Drs. M. J. Saraiva and P. P. Costa at Centro de Estudos de Paramiloidose (Portugal). Mapping of the gene was carried out in the laboratory of Dr. R. Sparkes (University of California at Los Angeles). Much valuable information on genetic screening for FAP was provided by Drs. M. Benson (Indiana Univ.), S. Ikeda (Shinshu Univ.) and M. J. Saraiva (Centro de Estudos de Paramiloidose).

Bibliography

Clinical and molecular basis of FAP may be obtained from the following recent reviews:
1) Benson, M. and Wallace, M. Amyloidosis. In: *The Metabolic Basis of Inherited Disease,* ed. C.R. Scriver et al. pp 2439-2460, McGraw-Hill, New York, 1989.
2) Saraiva, M. J. M. et al. Transthyretin (Prealbumin) in familial amyloidotic polyneuropathy: Genetics and functional aspect. *Adv. in Neurol. 48:* 189-200, 1988.
Molecular genetic analyses of FAP are summarized in the following review papers:
1) Sakaki, Y. et al. Genetic analysis of familial amyloidotic polyneuropathy, an autosomal dominant disease. *Clin. Chim. Acta 185:* 291-298 (1989).
2) Sakaki, Y. et al. Human transthyretin (prealbumin) gene and molecular genetics of familial amyloidotic polyneuropathy. *Mol. Biol. Med. 6:* 161-168 (1989).
Haplotype analysis of FAP is described in:
1) Yoshioka, K. et al. Haplotype analysis of familial amyloidotic polyneuropathy: an evidence for multiple origins of the Val→Met mutation most common to the disease. *Human Genet. 83:* 9-13 (1989).
(Analysis of European FAP is now in preparation.)
Prenatal diagnosis of FAP is reported by:
1) Almeida, M. R. et al. Prenatal diagnosis of familial amyloidotic polyneuropathy: evidence for an early expression of the associated transthyretin methionine 30. *Human Genet.* (in press).
2) Nichols, W. C. et al. Enzymatic amplification of prealbumin genomic sequences and potential use in diagnosis of hereditary amyloidosis. *Am J. Hum. Genet. 41:* A230 (1987).

AN INTERNATIONAL CODE OF ETHICS IN MEDICAL GENETICS BEFORE THE HUMAN GENOME IS MAPPED

John C. Fletcher*, with Dorothy C. Wertz**

In modern health care, duties grounded in an ethical principle of respect for persons[1] can be made real and carried out only in communities of persons where respect prevails. These duties include: a) a primary (but not absolute) duty to respect and protect the autonomous choices and privacy of competent persons, and b) a duty to respect and protect persons whose reasoning is impaired, who have permanently lost the capacity to participate in decision-making, or who have never had such capacity at all, e.g., seriously retarded or disabled persons, or patients in states of permanent unconsciousness. These impaired and incapacitated human beings are extremely vulnerable except when encompassed within communities of persons and the respect which is due to persons.

In modern societies marked by loss of communities in which to nurture persons, their relationships and their moral duties, the gap between the ideal — described by the duties — and actual practice is often very large. How can individuals who are essentially strangers to one another in "medical encounters" actually discover, learn and perform such duties? The pace of massive technological change exacerbates the gap, which brings us to the point of this paper.

When the human genome is mapped, what will become of these specific duties, viz., 1) respect for and protection of the autonomous choices and privacy of persons, and 2) respect for and protection of vulnerable human beings who are or become unable to express autonomous choices? Will such duties survive intact? Or will they be diminished and disappear for many of the most vulnerable?

These questions point to advances in genetic knowledge, advances that will gradually affect every person with or without the capacity to choose genetic screening and testing. Information about the molecular contents of virtually every gene harmful to human health can be — and gradually will be — used for testing. Human beings will be genetically laid bare and vulnerable as never before. Approaches to diagnosis, treatment and prevention of genetic and multifactorial diseases will grow and proliferate. In such a context, what ethical guidance will be needed to respect and protect the autonomous choices of persons — and those who have no capacity to choose? What laws will be needed?

The medical specialty most affected by these changes is medical genetics. How will medical genetics be practised when the human genome is map-

---

*   Professor of Biomedical Ethics and Religious Studies, University of Virginia, Charlottesville, Virginia, U.S.A.
** Research Professor, Health Services Section, School of Public Health, Boston University, Boston MA, U.S.A.

ped? This specialty involves research, diagnosis and treatment concerning the relation of heredity to disease. Genetic services now include screening (carriers and newborns), genetic counselling, prenatal and postnatal diagnosis, and treatment; among various approaches to treatment, human gene therapy should be expected. Medical genetics is where genetic knowledge is most frequently applied to human problems today. Medical geneticists are also teachers and mediators to other biomedical fields concerning their experience with ethical problems in human genetics.

Many are concerned lest genetic knowledge will be abused by employers and in the workplace, or by insurers, or even in the health care system. Given the evidence of discrimination in the United States against persons with HIV infection, cancer and other disabling disorders [School Board of Nassau County v. Arline, 480 US 273 (1987)], society will likely not trust geneticists and other scientists alone to use responsibly the power which genetic knowledge gives them, in the workplace or educational setting. The dangers of isolation, and of loss of insurance, educational and job opportunities, to persons diagnosed with incurable and costly disorders known from early childhood are real to many who are dubious about clinical uses of the "new genetics." What can be done? Can there be enough shared agreement about ethical guidelines for the practice of human genetics in the future? Society needs reassurance that these questions can be answered positively. Society can also stop the genome project altogether — by cutting off funds or by legal action — if convincing and reliable ethical and legal guidelines are needed for genetic testing?

Our paper responds to these questions and proposes that it is timely, nationally and internationally, for medical geneticists and their colleagues in allied fields (such as obstetrics and gynecology, pediatrics and internal medicine) to debate the features of a code of ethics for human genetics. We argue that major attention to ethics and law, and to strengthening communities of human geneticists, is needed to help a transition from the "old" to the "new" genetics.

Social futures: links with prophetism

Deeper resources in religion and philosophy, beyond ethics and law, will also be needed. What kinds of communities and persons are needed to sustain and even enrich a transition from the "old genetics" to a "new genetics" in which eventually every gene that harms or helps human beings will be mapped and gradually sequenced? What kinds of social futures do we envision for the context of a new genetics?

Concern about social futures and about renewal of human communities is associated with a religious tradition of *prophecy*. Prophetism is one resource, among the world's religious traditions, for the challenges of human genetic advances. Paul Tillich[2] wrote that the vocation of prophetism must not be unduly restricted to the Hebrew prophets, despite their uniqueness. He noted that the common marks of prophetism in religious history were their "attack on a given sacramental system" and emphasis on "fulfilment in the future" rather than in the present, which mysticism envisions. Prophetism is marked, then, by a) an ultimate source of *judgment* from within and beyond any specific culture and all of its projects, and b) a *promise* of future fulfilment, with signs of fulfilment in the present. In weighing great historical events, such as the human genome mapping project, a prophetic view aims at the opening of minds by judgments alongside promises. Practical reasoning needs judgment and renewal to remain sound. Toulmin, a philosopher, strikes similar themes when he writes that human beings show rationality, including practical ethical reasoning:

> not by ordering their concepts and beliefs in tidy, formal structures, but by their preparedness to respond to novel situations with open minds — acknowledging the shortcomings of their former procedures and moving beyond them.[3]

History casts doubt on whether human reason has inherently within it the power to "open minds" and keep them open, as well as the power to "acknowledge shortcomings and move beyond them." Modern societies have many and diverse sources of judgment and promise for ethical and legal guidance. We point to a prophetic context of judgment and promise as one of those sources.

Are there moral dangers in a "new genetics?" In developed nations, we see the dangers as: 1) a non-ethos of chaos, with no common moral agreements at all about how genetic knowledge should be used; 2) moral polarization between practices of excessive paternalism or individualism; or 3) societal coercion of persons with genetic disorders or at higher risk of transmitting them, in the name of cost/benefit considerations or a eugenic vision. Data that we report below show that it is unlikely that these dangers will arise from geneticists' practices.[4] Ironically, an excessive concern to protect autonomy could result in a social "backlash" that would reduce support for genetic services. An absolutist approach to autonomy never judges the motives of patients or the consequences of their actions. In short, autonomy absolutists never say "No." If, in the name of autonomy, patients in genetic services were given anything they wanted, e.g., gender selection or fetal tissue typing to find suitable donors for transplants after birth, society might react punitively and reduce access to genetic services.

What are the ethically acceptable exceptions to respect for autonomous choices? In the final section, we recommend some basic approaches to major ethical problems in human genetics, which could, after appropriate debate and discussion by geneticists, become the core of a code of ethics for medical geneticists. If such practices were adopted in a code and followed, the chances of a backlash would be reduced.

Threats could arise also against diagnosed persons from interested third parties who become aggressive, e.g., insurers, employers and government agencies. The desire to contain the lifelong cost of genetic diseases may eventually dominate the security and privacy of those at higher genetic risk, unless laws and policies to prevent or punish "genetic discrimination" are enacted. Laws against genetic discrimination are clearly necessary. The Americans with Disabilities Act, approved by the U.S. House of Representatives and now under debate in the Senate[5] will also be discussed in the final section.

Ethical issues in human genetics must be addressed by national and international societies of geneticists and colleagues in allied fields. The genome mapping project is international in scope. Scientists work together in international teams and see patients from different cultures. We are searching for a common language to understand and approach ethical problems that arise in the use of genetic knowledge. This language is designed to identify, analyse and attempt to resolve conflicts that arise largely between the duties involved in respect for persons and other duties that enjoin protection of others from harm in communities of persons.

Four premises — duly qualified

We begin from four premises to scan the various forms of social future that will likely be the context of a new genetics. To conserve space, each premise is only partly supported by arguments to justify it. Each premise also has one or more qualifications to render it plausible and debatable. In this paper, we can discuss only the fourth premise in depth. In future work, we shall enlarge upon the first three premises in relation to the fourth.

1. The Human Genome Project will succeed

Our first premise, based on the progress of science, is that the estimated 50,000 to 100,000 genes that comprise the genome will be mapped successfully through combined international projects. Many of these genes will also be sequenced. Other speakers will describe the scientific aspects of the human genome project in relation to screening and therapy. It is important to remember, however, that the genes that will be mapped and sequenced will not come from one human being, but from cell lines that have been acquired and grown in many laboratories over a long period. Many people are unaware of this important fact, and they mistakenly believe that the genes of only one person will be mapped and sequenced.

"Mapping," a process using family-linkage studies and biochemical measurements, results in knowing the location of a gene on a chromosome. "Sequencing" means the breaking down of the biochemical parts of DNA, the large molecule that composes the gene, into its parts, called nucleotides. Where does the human genome project stand today? Approximately 1900 (0.02% — 0.04%) genes have now been mapped to their loci on the chromosomes. New techniques will hasten further the sharp increase of gene mapping in recent years. Newer techniques called "chomosome jumping" permit much faster results in mapping. The gene for cystic fibrosis was mapped in about four years with a jumping technique, instead of the

18 years or so that "walking" would have taken.[6] Twenty years ago Wu and Taylor sequenced the first piece of DNA, using the most painstaking methods, by hand[7]. Automated methods of sequencing by the use of laser beams to read the order of the base molecules in purified samples of DNA speed up this process tremendously. Years of work by hand are reduced to weeks by automated sequencers.

The first premise should be qualified in two ways. First, in the U.S., delays may be caused by opposition of non-geneticists whose funding is reduced by the costs of the genome project. The National Research Council of the U.S. National Academy of Sciences has estimated these costs to be at least $3 billion by the year 2010, and forecast in 1988 that the mapping project would be completed in 15 years[8]. If five years are added to allow for controversy and delay, the mapping aspect of the genome project should be completed by 2010. Every gene or combination of genes involved in serious genetic disorders will be mapped. The DNA for many genes will be sequenced, which means that opportunities to develop drugs or treatment at the molecular level will be vastly multiplied.

However, one cannot be confident that such scientific knowledge will be readily applied to the diagnosis and treatment of genetic diseases, since in the U.S. real obstacles exist to federally-supported applied research. These obstacles could arise also in other nations, especially in Germany, where a law against any embryo research is pending[9]. Such obstacles pose a second qualification. Even if the human genome project were completed today, in the U.S. federally-funded applied research into fetal diagnosis and genetic therapy are blocked by restrictions or moratoria on federal funding of research involving human embryos, the fetus and fetal tissue[10,11]. The reason is that such research activities are associated with elective abortion or the death of the embryo in the context of research. Groups with strong moral views see this association as unethical. Consequently, the U.S. faces a dire contradiction. Great sums will be spent on mapping and sequencing the human genome, but the federal government will be prohibited from funding clinical research based on the findings of the genome project if such research involves studying genes in human embryos, prenatal diagnosis of early embryos for genetic testing, attempts to manipulate genes in embryos (fetal gene therapy), or attempts to transplant altered or unaltered fetal genes into humans with genetic diseases. Such research is either prohibited or so restricted that it cannot be done with federal funds. The options are: reduce the obstacles to federal funding for clinical applications of the research, or stop federal funding of the genome project because of moral objections to the uses of the knowledge, or fund applied research only from private sources (resulting in no federal regulation in these vital areas of research ethics!). These obstacles need to be negotiated into creative compromises in social policy to permit clinical research to proceed. Parliament in the United Kingdom recently passed an Embryo Research Act, which, coupled with the country's fetal research policy, permits carefully restricted research in these areas[12].

2. Genetic services will become part of universal health care

Our second premise is that genetic services will gradually be a part of national health plans — supported by universal health insurance with other added insurance options — in all nations with universal health planning and insurance. Today, the United States lacks a universal approach to insurance for basic health care. It is likely that society will empower leaders in government, industry and labour in the United States to develop a basic national health plan and universal health insurance for all of its people and the strangers in its midst. In any health care system, including a reformed U.S. system, genetic services (screening, counselling, prenatal diagnosis, and treatment for genetic diseases) will be included in a basic "floor of services" to be available to all identified as being at higher genetic risk or affected by a genetic disorder.

The issue of *which genetic services and how much* to include in a national health plan is both ethical and political in nature. We believe that the American people will eventually empower leaders to effect a national health plan. The contradictions and costs that are permitted in the present arrangements will become too unbearable to persist into the 21st century. In our view, the moral case for a universal health plan grounded in distributive justice has been made in a broad body of literature and commentary[13-20]. The moral case for judgment is clear. The burden of proof is on those who oppose a national health plan. As contradictions accumulate, new power to effect change will be given to elected and other leaders.

Beyond the imperatives of justice, the success of the human genome project and screening for genetic diseases will also ease the way in the U.S. to national health insurance and a national health plan that will include human genetics. Why? Many more Americans will understand for the first time that they are not "to blame" for cancer, diabetes, heart disease or other common disorders. Genetics is a great equalizer, and eventually all will understand that they suffer from diseases and burdens with a strong to moderate genetic determinant. The potential exploitation by insurers of those at higher genetic risk for disease will force the nation to protect persons at higher risk by having a sufficient pool of social insurance to bear these risks.

A qualification is that progress to a U.S. health plan may take many years, and controversies will arise about the inclusion of genetic services in the plan because of their association with abortion. For example, health planners in the Bush Administration envision only a modicum of genetic services in the national health plan by the year 2000, e.g., screening maternal serum for alpha-fetoprotein and newborn screening. Any references to genetic screening of carriers, prenatal diagnosis, abortion, or therapy at the genetic level are not in the planning document *[Promoting/Preventing Disease: Year 2000. Objectives for the Nation.* Public Health Service, Bureau of Maternal and Child Health, 1990, draft memorandum].

Today the U.S. is the only developed nation whose elected leaders, because of moral opposition to abortion, plan to reduce genetic services. Americans will eventually decide — in their own enlightened self-interest — to chart the course of the nation's health in more moderate ways, inclusive of genetic services and other forms of health care grounded in preventive medicine and public health. Much can be learned in the U.S. about the role of genetic services in the total health care systems of other nations, which is now well described in 18 other nations, including our host nation, Japan[4].

3. Genetic knowledge will become part of everyday life

The third premise is that "genetic knowledge" will gradually become a normal part of everyday life, because genetic information will gradually transform the practice of medicine, and every physician will learn approaches to genetic diagnosis, prevention and treatment. In time, children will be raised with the intuition that it is "good to want to know" about oneself genetically to prevent *significant* harm to oneself, one's children and future generations. The completion of the human genome project will provide a basis for acting on a moral obligation to *future* generations, a claim that has appeared weak in the past. A generation *with* such knowledge who neglected to use it to minimize the genetic risks in reproduction could hardly be said to respect the requirements of intergenerational justice. John Rawls, in a noted treatise on justice[21], written well before the DNA revolution, saw the use of genetic knowledge to prevent the most serious genetic defects as a matter of intergenerational justice. He argued that such actions are fair in terms of what the present generation, whose genetic inheritance is fixed, owes to later generations.

Persons are less likely to "want to know" about themselves genetically if they are afraid that such knowledge will be abused or used to punish them. Our premise assumes that society will develop ethical and legal means to protect higher-risk persons from genetic discrimination and stigmatization. Universal access to genetic services will also be available to act on the perception that it is "good to want to know" about genetic risks. Genetic information will become less threatening and stigmatizing when the benefits of diagnosis and treatment are more evident.

4. The practice of medical genetics is already subject to ethical agreements

Practice and the use of knowledge gained by geneticists, employers and insurers will need reliable ethical guidance and legal protection. Our fourth premise is that the core of a set of ethical agreements about showing respect for autonomous choices of persons and controlling abuses of genetic knowledge is now present in the practice of medical genetics in many nations. If this premise is true, there is no need to search frantically for a "new ethics" for the "new genetics."

103

There are no new ethical problems, in the sense of problems not seen before, in the new genetics. The eight ethical problems in medical genetics described below arose with amniocentesis and carrier screening in the late 1960s. However, these problems will be magnified in complexity and frequency in the "new genetics." Holtzman's work describes this magnification process and its consequences for medical genetics and public policy[22].

However, a major qualification is needed at this point. Professional groups of medical geneticists, national or international, have hesitated to frame *in a written code* their agreements about common ethical approaches to problems. Their agreements are in an oral tradition or implied in their literature. This conservative stance may be due to the newness of their specialty. Perhaps medical geneticists are not sufficiently yet a *professional community* with clear traditions to feel confident about setting forth their ethical standards. In our view, geneticists persist with an oral tradition in ethics to the detriment of their specialty. Four reasons lie behind this judgment: 1) those with the greatest power in human genetics today — the geneticists — are thereby less accountable to the public; 2) geneticists should expect their own moral views to count in shaping future public policy, which will be more effective after moral commitments reached after debate and decision; 3) geneticists in this generation are obliged to transmit their moral commitments to those who follow them; and 4) the marks of a profession include a code of ethics. Are geneticists "sacramentalizing" their roles in an oral, priestly tradition that can be openly examined and criticized by others?

Other concerns that might make geneticists reluctant to have a code may be: a) ethics seems too subjective or personal to commit to writing; b) genetic technologies change too fast; c) legal concerns, e.g., that lawsuits may be brought against geneticists who do not follow guidance from a code; or d) beliefs in ethical relativism may suggest that a written code will repress important minority views and lead to domination by Western interests. Each concern can be met with a reasoned answer. Many scientific and professional groups have written codes of ethics. Ethical guidance should not depend too heavily on trends in technology, although what ought to be done cannot be divorced from what can be done. Professional ethical codes are not laws. Also, codes of ethics cannot foresee every situation that will arise or define every ethical action that might be taken. If actions that appear to be beyond the scope of a code must be taken, as long as these are soundly argued and ethically grounded it is unlikely that any lawsuit based on a claim that a *code was not followed* would succeed. Lawsuits are won because patients are harmed and duties are neglected. The previous section on the relation of ethics and law is relevant just at this point.

Our fourth premise is that the substance of widely shared but unwritten moral agreements found in the practice of medical genetics today is sufficient, with three added exceptions, for national or international proposals on approaches to ethical problems among geneticists. Our task is to point to areas of agreement and disagreement, with proposals to modify the latter. Geneticists are more of a community — nationally and internationally — than they may have imagined themselves to be. Geneticists themselves may be crucial agents of moral change to adopt approaches that need not be rigid or legalistic in substance. A code of ethics can be a resource, among others, from which to continue to shape ethical and public policy responses to the "new genetics."

Eight ethical problems in medical genetics

Geneticists and their patients in a cross-cultural setting face a common set of eight ethical problems. These are ranked in order of significance according to three factors: 1) the results of an international survey, 2) frequency of discussion among geneticists and medical ethicists in 19 countries[4], and 3) numbers of persons whose welfare is adversely affected by the problem.

1. Fairness of access to genetic services

Geneticists around the world note that unfair or no access to services is the most widespread ethical problem in human genetics. The problem has two aspects: a) unfairness in access to genetic services, and b) insufficient services to meet needs. Studies of utilization of genetic services find that the problem particularly affects families and pregnant women whom physicians do not refer to genetic services, who suffer from poverty or lack of education, or who live at a distance from a genetic centre. Geneticists in many countries have expressed concern lest spread of private, for-profit clinics will give privileged access to patients who can pay out-of-pocket. In the U.S., genetic services mut be seen in the context of larger inequalities in prenatal care. An estimated 25% of all mothers, or 939,000 women annually, receive late or no prenatal care[23]. Women who obtain prenatal diagnosis are, disproportionately, white, well educated and well-off. If this trend continues, genetic handicap could become a mark of social class. Friedmann's recent article[24] indicates that services involving "reverse genetics" are likely to become available to the U.S. public slowly, and always inadequately to meet the true need. However, even countries with national health insurance experience some degree of social and geographical inequality in the use of genetic services.

2. Abortion choices

In the United States, and to varying degrees in other nations, abortion choices for genetic reasons present the second most significant area of moral conflict and concern[25]. Abortion choices are especially difficult when pregnancies are planned and wanted, as are many among couples who use counselling and prenatal diagnosis. Other factors that create conflict about abortion choices in pregnancies at some genetic risk are:
    a) beliefs about the higher moral status of the fetus in the middle

trimester, when a decision about abortion must be made after amniocentesis;

b) the wide spectrum of severity in some disorders, e.g., sickle-cell disease;

c) the treatability of some disorders, e.g., phenylketonuria, which can now be diagnosed prenatally by means of DNA techniques;

d) the guilt that parents feel about a living child of theirs with a disorder, when considering abortion of a newly diagnosed affected fetus;

e) concern lest the moral reasoning that justifies a practice of selective abortion establishes a precedent for neglect of persons with genetic disorders who are born and survive; other critics also compare the reasoning for abortion to arguments for active euthanasia in severely handicapped newborns;

f) decisions about abortion in twin pregnancies in which one fetus is healthy and the other affected;

g) a decision not to abort after a positive genetic finding, on the part of a woman or couple who may be subjected to pressure to abort or threatened with loss of medical care. Rothman[26] has discussed the social pressure felt by some women in the U.S. to use abortion after a prenatal diagnosis of a genetic disorder.

3. Problems of confidentiality

Difficulty in maintaining confidentiality in the patient-geneticist relationship is a significant problem everywhere. Geneticists have a duty to prevent unconsenting disclosure of their patients' genetic diagnoses and health prognoses. This duty can sometimes conflict with interests of relatives at risk[27].

4. Protecting privacy from institutional third parties

Geneticists see that persons at higher genetic risk are very vulnerable to the collective interests of institutional third parties, such as government agencies, health insurers and employers. Under what conditions, if any, should such third parties have access to personal genetic data?

5. Disclosure dilemmas

Various disclosure dilemmas are associated with the practice of medical genetics. Geneticists have access to psychologically sensitive information that can disrupt family and marital relations if communicated without careful thought and preparation. Such situations arise when, e.g.:

- the geneticist knows, but a married couple does not yet know, which parent has transmitted a disorder to a child;
- chromosome tests reveal that a phenotypical but infertile female has an XY genotype;

106

- tests reveal false paternity, and the identity of the biological father is a secret;
- the geneticist but not the husband is aware of previous elective abortions;
- geneticists disagree about the interpretation of laboratory findings and abortion might occur if the disagreement is disclosed;
- disclosure of a genetic diagnosis to a vulnerable or fragile individual may carry a risk of psychological harm.

6. Indications for prenatal diagnosis

Non-medical or non-genetic indications for prenatal diagnosis are ethically controversial, e.g., requests due to "maternal anxiety", in the absence of any family history of genetic disorders; prior refusal of abortion on the part of the woman or couple; and sex selection unrelated to any sex-linked disorders. However, such ethical problems affect the welfare of far fewer persons than do the problems described above.

7. Voluntary or mandatory screening

Whether genetic services, especially genetic screening, should be voluntary or mandatory has been an ethical issue with regard to sickle-cell carrier screening in the U.S.[28], and is a subject about which geneticists have a variety of views[29].

8. Counselling incapacitated persons

"Nondirective" genetic counselling means that counsellors refrain from giving moral advice, even if asked, out of respect for the moral autonomy of patients. Counsellors encourage the patient or parental couple to make moral choices in the light of their own values. This is the prevailing ethical approach to genetic counselling in many nations. However, ethical problems do arise in choices to be nondirective when counselling patients with impaired capacity to comprehend and participate in decision-making. They may be mentally ill or severely retarded, or abusers of alcohol or drugs. Some may be unable to communicate because of poor education. Patients from a different culture may differ from the geneticist in their views about science and nature. For these reasons patients may be unable to weigh the significance of genetic risks. These cases were infrequently mentioned by our international-survey respondents, but were the subject of notes on questionnaires pointing out that such problems exist, especially in cases of counselling severely retarded persons or substance abusers.

Any adequate set of approaches to ethical problems in medical genetics must begin by addressing these problems. The next section describes in some detail areas of agreement and disagreement among medical geneticists in 19 nations on five of the eight problems.

An International Study of Medical Geneticists

How do geneticists approach such ethical problems in their practice? Geneticists as individuals have discussed their own views[30-34]. However, no systematic study of approaches to such problems, especially in an international context, had ever been done. After a proposal[35], an action-research project was designed to study the degrees of consensus and variation among medical geneticists in 19 nations. Action research, in the words of its founders[36,37], is the use of psychological and social research to identify social problems in a group, coupled with active participation of the investigator and members of the group to understand and resolve these problems.

In field-work prior to the study, geneticists identified ethical problems in practice, which were described in 14 clinical cases, along with four questions on screening. Another part concerned questions about goals and approaches to counselling, originally proposed by Fraser[38], which replicated questions asked by previous researchers[39]. For the clinical cases and screening questions, respondents were asked, from a list of options, what they would do, and why, in their own words, they chose an option. They were also asked which cases they found most and least ethically difficult. We used a standard for "consensus" of 75% of the nations. Because the methods and results of this study have been published elsewhere[40], we briefly summarize its findings below.

Results

Cases with consensus about approaches

Respondent geneticists strongly share common approaches to 5 (36%) of the 14 clinical cases that concerned patient autonomy and protection of privacy:

Case 1: 98% would disclose conflicting findings
Case 2: 97% would disclose ambiguous laboratory results
Case 3: 94% would disclose unproven findings
Case 4: 96% would protect the mother's privacy
Case 5: 88% and 84% would counsel non-directively in these cases

Cases with less than strong consensus on approaches

Respondent geneticists had less than strong consensus on several cases that also involved reproductive freedom and autonomy, but the views of the majority strongly went in a clear direction:

Case 10/11: 83% favour presenting artificial insemination by donor (AID) as a reproductive option

Case 12: 85% favour prenatal diagnosis even if the couple refuses abortion in advance

Case 13: 73% favour prenatal diagnosis for maternal anxiety

108

Cases with least consensus on approaches

Geneticists shared least consensus on a number of cases involving sharp conflict of ethical principles and clashes of interests, which were very controversial:

Case 6: 58% would tell the relatives of the Huntington patient

Case 7: 60% would tell the relatives of the hemophilia A carrier

Case 8: 54% would disclose, unasked, which parent was a carrier of a translocation for Down's syndrome

Case 9: 51% would disclose to an XY female the reason for her infertility

Case 14: 58% would not cooperate in providing prenatal diagnosis for sex selection

Respondents in seven nations (Brazil, Federal Republic of Germany, Israel, Sweden, Switzerland, U.K. and U.S.A.) identified sex selection as posing the greatest ethical dilemma. In this case, a couple with four healthy daughters desire a son. They say that if the fetus is a female they will abort it, and if they are refused prenatal diagnosis they will abort it anyway, rather than risk having a fifth girl. Of the respondent geneticists, 24% (34% in the U.S.A.) would perform prenatal diagnosis solely for sex selection, and another 17% (28% in the U.S.A.) would offer a referral. Although a majority of geneticists in the U.S.A., Hungary and India would perform prenatal diagnosis for this couple or refer them to someone who would, they would do so for different reasons. American geneticists based their responses on respect for patient autonomy; Hungarian geneticists thought they were saving half the fetuses from abortion; and Indian geneticists spoke of limiting the population or preventing the suffering or early death of unwanted females.

Directive versus nondirective counselling

Most of the respondent geneticists subscribe to the ideal of nondirectiveness. When asked about the goals and conduct of counselling, 99.8% said that the first goal of counselling was to help patients understand genetic information so that they could make informed decisions; only 15% considered it appropriate to advise patients what to do.

Workplace screening

According to 72% of respondents, genetic screening in the workplace should be voluntary. Exceptions were those from Brazil, German Democratic Republic, Hungary, India and Turkey, where the majority considered it should be mandatory, as a means of protecting workers. Among these respondents there was strong consensus (81 to 89%) that employers and insurers should have no access to the results without the worker's consent, and 45% considered that insurers should have no access at all, even with consent. In all, 75% thought that, if and when population screening for cystic fibrosis became feasible, participation should be voluntary. When the survey was done, carrier screening for cystic fibrosis could not be done. The scientific situation has changed, but geneticists are guarded about extending such screening to whole populations[41].

Future priorities in ethics

All respondent geneticists ranked increased demand for services, carrier screening for common genetic disorders, and allocation of limited resources as the issues that should be of greatest ethical concern within the next 10 to 15 years. As less problematic they considered research on the human embryo, genetic screening in the workplace for susceptibility to occupationally related disease, long-range eugenic concerns and sex selection.

Implications of the study for approaches to a code of ethics

For lack of space, we do not include the qualitative aspects of the study of the moral reasoning of the respondents, which is reported elsewhere[4,42]. We believe that the results of this study support the claims made above for the fourth premise. There is strong agreement, worldwide, on some crucial approaches to several of the major ethical problems in the old and the new genetics. With regard to the problem areas on which consensus is less marked, many nations show strong trends that can be a basis for creative debate among geneticists about adoption of a code of ethics. However, if the practice of geneticists is similar to their responses to the study, patients around the world are not in moral danger from the practice of medical genetics. While allowance must always be made for overestimating the fit between the actual and the ideal, those most concerned about protection of autonomy and privacy by geneticists can surely be reassured by this study.

Some findings were a surprise. Despite controversies about new reproductive options, most respondents were willing to discuss donor egg and surrogate mothering in a non-directive context. Most will do prenatal diagnosis for maternal anxiety alone, except in some countries where prenatal diagnosis is rationed through national health insurance. Also, we expected to find stronger opposition to prenatal diagnosis for sex selection unrelated to X-linked disease. In the U.S.A., 34% of respondents would perform prenatal diagnosis and another 28% were willing to refer the couple to a physician who would. In Canada, 30% of respondents were willing to perform, and 17% to refer. These findings are surprising, for studies conducted in the U.S.A. in 1972-73[43] and in Canada in 1975[44] found only 1% (of 448) and 21% (of 149) of genetic counsellors, respectively, willing to meet this request. However, this issue clearly needs re-study in the context of several types of cases.

We were not surprised to find controversy about confidentiality dilemmas. Respondents ranked this as the issue with the second most difficult ethical conflict (behind sex selection). About one-third would not favour breaching confidentiality, about one-fourth would actively inform the relatives, and about one-third would inform relatives only if they asked.

Geneticists are wary of institutional third parties, i.e., insurers and employers, having access to results of genetic screening without the patient's consent. However, respondents from only 10 of 18 nations agreed strongly that genetic screening in the workplace should be voluntary. Obviously, the social and economic structure of nations strongly influences ethical views. As regards legal protection against genetic discrimination in employment, the Americans with Disabilities Act, discussed by Orentlicher in the context of genetic screening in industry[5], prohibits employers from discriminating on the basis of a person's disability that does not interfere with performance of a job, as well as the use of medical (including genetic) tests to detect disabilities, unless the testing is for information about fitness for job-related tasks. For example, "a railroad would be permitted to test the sight of its engineers, but would not be able to measure the serum glucose level of its accountants to screen for diabetes[5]." Orentlicher points out that it is unclear whether persons with a prospective risk of disease, not currently affected or disabled, would be protected. This weakness needs to be remedied to include genetic screening for disorders of later onset.

Proposed approaches to ethical problems in medical genetics

Findings of the international study strongly support some approaches to ethical problems in medical genetics. We pose these approaches below in relation to the eight problems cited above, to stimulate debate among geneticists in professional and international organizations.

Any set of ethical guidelines should be prefaced with positive obligations of geneticists to their patients[35]. In the light of claims of basic ethical principles along with needs of patients and families and responsibilities of geneticists, and exceptions in extraordinary situations to override parental autonomy and privacy, these approaches *can* be adopted in many nations as a basis for an international code of ethics:

1. Access to genetic services

Securing access to adequate genetic services is an ethical obligation of societies to persons and families as part of a general obligation to secure access to adequate health care. Equitable distribution of genetic services is owed first to those at higher genetic risk, whose need is greatest. Geneticists should do everything within their legitimate authority to ensure that patients who most need services and have the greatest opportunity to benefit receive such services.

2. Respect for and safeguarding of parental choices

Geneticists in all nations consider it essential to protect rights of parental choice. Medical geneticists should safeguard the options of parents in genetic services, including the options of abortion or to carry to term a fetus with a malformation or a genetic disorder.*

An exception to this duty should be made in decisions that involve information about the sex of the fetus, unrelated to genetic diseases, that may lead to abortion. Geneticists have no duty to cooperate with parental desires to abort for sex selection, because: 1) sex is not a genetic disease; 2) equality between males and females is violated; 3) sex selection is a precedent for eugenics, i.e., interventions in human reproduction with regard to characteristics unrelated to a genetic disease. Sex selection discredits the public image of prenatal diagnosis and of medical genetics.

3. Confidentiality when other family members are at high risk

Confidentiality is a vital but not absolute norm in medicine and in medical genetics. If the patient refuses to disclose proven risks of harm to relatives, the imperative of prevention of harm to others limits a physician's or counsellor's duty of confidentiality.

The President's Commission for the Study of Ethical Problems in Medicine recommends that confidentiality be breached only in exceptional circumstances that meet four conditions: (1) reasonable efforts to elicit voluntary consent to disclosure have failed; (2) there is a high probability both that harm will occur if the information is withheld and that the disclosed information will actually be used to avert harm; (3) the harm that identifiable individuals would suffer would be serious; and (4) appropriate precautions are taken to ensure that only the genetic information needed for diagnosis and/or treatment of the disease in question is disclosed[47].

Legal protections for geneticists who breach confidentiality to warn others at high risk, after having complied with these conditions, may be necessary in the future.

4. Protection of patients' privacy from institutional third parties

Without effective legal protection from genetic discrimination in employment, employers should not have access to personal genetic

---

* An important moral line about selective abortion exists in the public mind in the U.S.A. As put by Elias and Annas: "as controversial as elective abortion is, the use of abortion when a woman is carrying a fetus with a *severe* (italics added) genetic defect is very well accepted by a vast majority of both the public and physicians[27].

Data collected by the National Opinion Research Center[45] between 1972 and 1987 on trends in U.S. public opinion on genetic reasons for abortion show that between 75% and 77% support abortion for a serious genetic disease. The U.S. public's views also overwhelmingly support (89%) making genetic testing available for serious and fatal genetic diseases[46]. In the wake of the U.S. Supreme Court decision *Webster v. Reproductive Health Services*, state laws on abortion practices need to protect the choices of parents after prenatal diagnosis.

data. Even when legal protections have been adopted, access to personal genetic data should occur only with the informed consent of the individual involved. In nations where private health insurance is a major industry and third-party payer for health care, the industry should be regulated by government to discourage excessively costly premiums for those at known higher genetic risk. Refusal of private health insurance to families at higher risk for genetic disease should be illegal. Another alternative is government underwriting of health insurance for those at genetic risk.

5. *Full disclosure of all clinically relevant information*

Medical geneticists should disclose all clinically relevant information to patients and family members, consistent with considerations of immaturity and psychological well-being. Also, the capacity of those counselled to comprehend information and make voluntary choices are morally relevant factors in disclosure dilemmas. Psychological assessments of capacity are best done in consultation with mental health professionals. Psychologically sensitive information, such as XY genotype in a female, should be disclosed only when full and supportive counselling and patient education are assured.

6. *Use of prenatal diagnosis*

Prenatal diagnosis should be used only to give parents and physicians information about the health of the fetus. Any other use, such as for sex selection (except for X-linked disease) or for tissue-typing to plan for transplantation to benefit another person after birth, should be avoided. If patients have a genetic reason for diagnosis and also show excessive interest in the gender of the fetus, geneticists can consider delayed disclosure of gender after timely disclosure of clinical findings (M. Hultgren, personal communication).

Laws prohibiting abortion for sex selection are appropriate only where there is evidence that the medical indications for prenatal diagnosis are abused[48]. Where they are not abused, laws prohibiting sex-selection abortions not only are unneeded but also set harmful precedents restricting abortion choices.

7. *Voluntary approach to genetics services: one exception*

Genetic services programmes should be voluntary, with one exception, viz., screening newborns when treatment is available for those affected by genetic disorders. Forcing persons to submit to genetic services to prevent genetic disorder violates the principle of respect for persons and disrupts their relationships in families. Coercion, arbitrariness and discrimination against persons at higher genetic risk should be illegal in the context of employment and educational opportunities.

113

8. Non-directive counselling, with one exception that requires further study

Non-directive counselling, i.e., to refrain from direct moral advice in order to protect and enhance the autonomous choices of patients, is a commitment of medical geneticists, assuming that those being counselled know all the relevant facts, and that efforts are made to encourage them to consider the facts in the context of their beliefs and values.

A possible exception to non-directive counselling can arise in genetic counselling with incapacitated patients, especially when genetic harm to others is a clear and present danger. The incidence of this type of situation and geneticists' response to it need careful study. In principle, giving direct moral advice to relatives of incapacitated clients or to impaired clients themselves is ethically acceptable, if the likelihood of harm to others is great and if the geneticist has informed the patient and/or relatives in advance of counselling that "directive counselling" may be indicated. Geneticists should follow established protocols in determining the capacity of their patients[49].

Summary

These data show that medical geneticists in many countries strongly desire to respect the autonomous choices and the privacy of the persons they serve — with three, admittedly controversial, exceptions, viz., requests for prenatal diagnosis unrelated to the health of the fetus, confidentiality dilemmas when identified members of a family are at high genetic risk, and non-directive counselling when patients are impaired or incapacitated. In all other circumstances, respondents overwhelmingly supported the standard of "non-directive" genetic counselling, i.e., refraining from giving direct moral advice.

To return to the paper's beginning, duties grounded in ethical principles can be made real only in communities of persons. Advances in human genetics hold great promise of relief from the suffering of burdens of genetic disease along with moral dangers noted at many points above.The time has come for geneticists to contribute directly to the moral evolution of their field. Geneticists have the greatest experience with ethical problems in their field and need to define clear ethical guidelines for themselves as professionals and for their patients. They have a special responsibility to lead the way into and through a new era in genetics, not only as scientists but also as responsible practitioners. Except for more data on non-directive counselling in cases of impaired clients, the time for study is past. Professional codes of ethics are the best first step in the evolution of ethics and law to prepare for a future when the human genome is mapped.

Acknowledgements

Research for the international study was supported by the Medical Trust, one of the Pew Memorial Trusts, administered by the Glenmede Trust Company, Philadelphia, PA.; the Muriel and Maurice Miller Foundation; and the Norwegian Marshall Fund. Support was also given by the Warren G. Magnuson Clinical Center of the National Institutes of Health.

References

1. Beauchamp, T.L. & Childress, J.F.: 1989, *Principles of Biomedical Ethics*, 3rd ed. Oxford University Press, New York.
2. Tillich, P.: 1950, *Systematic Theology*, Vol. 1, University of Chicago, Chicago, pp. 141-4.
3. Toulmin, S.: 1972, *Human Understanding*, Princeton University Press, Princeton, NJ, vii.
4. Wertz, D.C., & Fletcher, J.C.: 1989, *Ethics and Human Genetics: A Cross-Cultural Perspective*, Springer-Verlag, Heidelberg.
5. Orentlicher, D.: 1990, Genetic screening by employers, *Journal of the American Medical Association* 263: 1005-8.
6. Rommens J.M. *et al.*: 1989, Identification of the cystic fibrosis gene: chromosome walking and jumping. *Science* 245: 1059-65.
7. Wu, R. & Taylor, E.: 1971, Nucleotide sequence analysis of DNA. II. Complete nucleotide sequence of the cohesive ends of bacteriophage lambda DNA. *Journal of Molecular Biology* 57: 491-511.
8. National Research Council, Committee on Mapping and Sequencing the Human Genome: 1988, *Mapping and Sequencing the Human Genome*, National Academy Press, Washington, D.C., p. 90.
9. Embryo research: 1989, *Institute of Medical Ethics Bulletin*, No. 48, March, pp. 3-4.
10. Fletcher, J.C. & Ryan, K.J.: 1987, Federal regulations for fetal research: A case for reform. *Law, Medicine, and Health Care* 15: 126-38.
11. Association of American Medical Colleges: 1988, *Fetal Research and Fetal Tissue Research*, AAMC, One Dupont Circle, N.W., Washington, D.C. 20036.
12. Embryo research. Britain votes yes. *Nature* 344: 799 (1990).
13. Outka, G.: 1974, Social justice and equal access to health care. *Journal of Religious Ethics*, p. 24.
14. Daniels, N.: 1985, *Just Health Care*, Cambridge University Press, Cambridge, MA.
15. The President's Commission for the Study of Ethical Problems in Medicine and Biomedical and Behavioral Research: 1983, *Securing Access to Health Care*, Vol. I, U.S. Government Printing Office, Washington, D.C.
16. Wikler, D.: 1983, Philosophical perspective on access to health care: an introduction, In: The President's Commission for the Study of Ethical Problems in Medicine and Biomedical and Behavioral Research, *Securing Access to Health Care*, Vol. II, U.S. Government Printing Office, Washington, D.C., pp. 107-52.
17. Dougherty, C.J.: 1988, *American Health Care. Realities, Rights and Reforms*, Oxford University Press, New York.
18. Callahan, D.: 1987, *Setting Limits: Medical Goals in an Aging Society*, Simon & Schuster, New York.
19. Callahan, D.: 1990, *What Kind of Life? The Limits of Medical Progress*, Simon & Schuster, New York.
20. Fleck, L.M.: 1989, Just health care (I): Is beneficence enough? *Theoretical Medicine* 10: 167-82; Just health care (II): Is equality too much? *Theoretical Medicine* 10: 301-10.
21. Rawls, J.: 1971, *A Theory of Justice*, Belknap Press, Cambridge, MA, pp. 107-8.
22. Holtzman, N.A.: 1989, *Proceed with Caution*, Johns Hopkins University Press, Baltimore, MD.
23. U.S. Department of Health and Human Services: 1988, *Health, United States, 1988*. DHHS Publication PHS 89-1232. Hyattsville, MD.
24. Friedmann, T.: 1990, Opinion: The human genome project — some implications of extensive "reverse genetic" medicine. *American Journal of Human Genetics* 46: 407-14.
25. Fletcher, J.C.: 1986, Moral problems and ethical guidance in prenatal diagnosis. Past, present, and future, In: Milunsky, A. (ed.), *Genetic Disorders and the Fetus*, Plenum, New York, pp. 823-38.

26. Rothman, B.K.: 1986, *The Tentative Pregnancy*, Viking Penguin Inc., New York.
27. Elias, S. & Annas, G.J.: 1987, *Reproductive Genetics and the Law*, Yearbook Publishers, Chicago, pp. 48-9.
28. Reilly, P.: 1973, Sickle cell anemia legislation. *Journal of Legal Medicine* 1: 4-39.
29. Wertz, D.C. & Fletcher, J.C.: 1989, An international survey of attitudes of medical geneticists toward mass screening and access to results. *Public Health Reports* 104: 35-44.
30. Berg, K.: 1983, Ethical problems arising from research progress in medical genetics, In: Berg, K., Tranoy, K.E. (eds.), *Research Ethics*. New York, Alan R. Liss.
31. Czeizel, A.: 1989, *The Right to be Born Healthy: Ethical Problems of Human Genetics in Hungary*. New York, Alan R. Liss.
32. Crawfurd, M. d'A.: 1983, Ethical and legal aspects of early prenatal diagnosis, *British Medical Bulletin* 39: 310-14.
33. Pfeiffer, R.A. *et al.*: 1982, Le généticien confronté aux problèmes d'éthique médicale, IXèmes Journées Européennes de Conseil Génétique, Erlangen, Sept. 1982. *Journal Génétique Humaine* 30 suppl.: 447-66.
34. Schroeder-Kurth, T.M.: 1982, Ethische Probleme bei Genetischer Beratung in der Schwangerschaft. *Monatsschrift für Kinderheilkunde* 130: 71-4; 1988, Ethische Uberlegungen zur pranatalen diagnostik. *Gynakologe* 21: 168-73.
35. Fletcher, J.C. *et al.*: 1985, Ethical aspects of medical genetics. *Clinical Genetics* 27: 199-205.
36. Lewin, K.: 1947, Group decision and social change. In: Newcomb T.M., Hartley E.L. (eds.) *Readings in Social Psychology*, New York, Holt, Rinehard, Winston, pp. 76-103.
37. Sanford, N.: 1970, Whatever happened to action research? *Journal of Social Issues* 26: 3-23.
38. Fraser, F.C.: 1974, Genetic counselling, *American Journal of Human Genetics* 26: 636-59.
39. Sorenson, J.R. *et al.*: 1981, *Reproductive Pasts, Reproductive Futures: Genetic Counselling and Its Effectiveness*. Alan R. Liss, New York.
40. Wertz, D.C. & Fletcher, J.C.: 1990, Medical geneticists confront ethical dilemmas: Cross-cultural comparisons among 18 nations. *American Journal of Human Genetics* 46: 1200-13.
41. Caskey, C.T. *et al.*: 1990, The American Society of Human Genetics Statement on Cystic Fibrosis Screening. *American Journal of Human Genetics* 46: 393.
42. Wertz, D.C. & Fletcher, J.C.: 1989, Moral reasoning among medical geneticists in 18 nations. *Theoretical Medicine* 10: 123-38.
43. Sorenson, J.R.: 1976, From social movement to clinical medicine: The role of law and the medical profession in regulating applied human genetics, In: Milunsky, A., Annas, G.J. (eds.), *Genetics and the Law I*. Plenum, New York, pp. 467-85.
44. Fraser, F.C. & Pressor, C.: 1977, Attitudes of counsellors in relation to prenatal sex determination for choice of sex, In: Lubs, H.A., DelaCruz, F. (eds.), *Genetic Counselling*. Raven, New York, pp. 109-20.
45. National Opinion Research Center, University of Chicago: 1987, *General Social Surveys, 1972-1987: Cumulative Codebook*. National Opinion Research Center, Chicago.
46. U.S. Congress, Office of Technology Assessment: 1987, *New Developments in Biotechnology*. Background paper: Public perceptions of biotechnology. OTA-BP-BA-45. U.S. Government Printing Office, Washington, D.C., pp. 74-5.
47. President's Commission for the Study of Ethical Problems in Medicine and Biomedical and Behavioral Research: 1983, *Screening and Counselling for Genetic Conditions*. U.S. Government Printing Office, Washington, D.C., p. 44.
48. Wertz, D.C. & Fletcher, J.C.: 1989, Fatal knowledge? Prenatal diagnosis and sex selection. *Hastings Center Report* 19: 21-7, May/June.
49. Applebaum P.S. & Grisso, T.: 1988, Assessing patients' capacities to consent to treatment. *New England Journal of Medicine* 319: 1635-8.

GENETIC SCREENING — POLICYMAKING ASPECTS

K. Takagi*

I am very glad to have a chance to talk on genetic screening. I am at present involved in national policymaking as a member of the House of Councillors of Japan. However, I was concerned with medical education and administration from 1935 to 1980, while devoting myself to research work as a physiologist. Therefore, I consider myself as a scientist rather than a Congressman, as my career as a Congressman has been only ten years long. Nonetheless, I have been interested in the subject of genetic diseases, with which are associated problems similar to those associated with brain death and organ transplantation, in which I have also been involved both as a biologist and as a policymaker, inside and outside of the Congress, from the viewpoints of ethics of life, definition of death, and its legalization, etc.

However, as I am a physician and biologist, not a specialist in genetics, I find it difficult to summarize within a limited time the issue of genetics, which covers a vast field and is making rapid progress. Therefore, please allow me only to introduce the present status of genetic screening in Japan with my personal views.

The issue of genetic screening

I understand that this meeting organized by CIOMS is one of a series that has been conducted for several years in line with the World Health Organization's strategy for attaining the goal of Health for All by the year 2000. I respect the fact that experts from all over the world have been intently discussing various issues from various viewpoints, with fruitful results. I assume that the issue of genetic screening was selected as a theme of this meeting also in this context. Thanks to progress in medicine and the efforts of many people, WHO declared the eradication of smallpox several years ago. It is very satisfactory that chronic diseases such as tuberculosis and leprosy as well as acute infectious diseases are coming under control, owing to improvements in environmental hygiene, nutrition and educational level. However, it is regrettable that thousands of people in the developing countries cannot receive the benefits of these improvements. In those countries, it is a very important issue, requiring the collaboration of other countries from now on, how to solve or reduce problems of social infrastructure by improving such aspects of it as basic education, literacy and the economy, which support medicine.

---

* Late Member of House of Councillors, Diet of Japan, Tokyo. (We regret the death of Dr Takagi shortly after the conference. Eds.)

The achievements of medicine in the developed countries owe much to thorough implementation of preventive measures against diseases, discoveries of new chemical and biological pharmaceutical products and progress in surgery, immunology and other specialties. At the same time, we should not overlook the fact that the environment, where all biological creatures including human beings live, is being destroyed and polluted by the rapid progress of science, upon which progress in medicine is also founded. As people realize the side-effects of pharmaceutical products and that Western medicine is not necessarily almighty, they began to turn their eyes to traditional Oriental medicine. Also, we are entering the age when human beings manipulate themselves at molecular level, wandering in the micro-world of molecular medicine. Political intervention into these matters is being urgently demanded.

While we are distracted by the unveiling side-effects of science-oriented culture on human beings, the threat of the "population problem" is insidiously but steadily growing. This topic was dealt with at a CIOMS conference two years ago under the theme of "Ethics and human values in family planning". The problem of family planning is one of "quantity", while the problem of genetics is one of "quality". I understand that I am expected as a policy-maker to give my comments on improvement of quality by means of genetic screening, and on genetic screening itself.

Unlike infectious diseases caused by foreign pathological agents such as viruses and bacteria, the pathological causes of genetic diseases reside inside the human being; genetic diseases are the diseases caused by errors or defects which are deeply rooted in DNA in the process of the development and evolution of the human being. Following Mendel's law of heredity or through mutations, genetic diseases are characterized by the fact that these errors or defects have been passed on from one generation to another without natural spontaneous cure. Nowadays, more than 4000 genetic diseases are known and every disease has its own almost fixed incidence rate. Since there is no natural cure, it is nothing but a misfortune for patients to have such genetic diseases, and it is a psychological and economic burden for families of patients. The purpose of genetic medicine is to reduce as much as possible, if possible to zero, the incidence of groups of people unfitted to the natural and social environment; relatively, this is a very minor group. This purpose can be classified into: 1) preventive measures at the embryonic stage or before marriage; 2) palliative therapy, which alleviates the damage caused to a newborn with a congenital impairment, permitting a more or less normal life; and 3) fundamental therapy, which manipulates DNA at the newborn or embryonic stage. When people hear the term "genetic screening", they may have an impression that compulsory administrative measures of genetic examination will be conducted on everyone at any stage of growth of the human being. If you recall the chaos that was caused by the compulsory screening for sickle-cell anemia in the United States of America in 1974, although its purpose was genuine, you can easily understand that compulsory screening would be unjustified.

There is no need to mention the inhumane conduct of the Nazis during World War II. This was a good lession for us to understand that genetic screening should not be compulsory at any time. In Japan, a screening test for colour vision for all schoolchildren has long been conducted in a semi-compulsory manner at school. As a result of this screening, it is reported that some schools restricted some of the colour-blind from the chance of higher education and from free choice of subjects, and that some companies in their employment policies discriminate against the colour-blind. Also, there have been some cases in which consanguineous marriage among patients was restricted. For these reasons, the Japan Ophthalmologists Association has requested the government to abandon screening for colour vision at school. The Association is also requesting that primary-school textbooks should make no mention of the heredity of colour-blindness as an example of X-linked recessive heredity.

All the medical and anthropological geneticists are against the idea of implementing compulsory and administrative genetic screening in Japan. In the U.S.A., after the bitter experience of screening for sickle-cell anemia, the National Genetic Diseases Law was passed in 1976. It stipulates voluntary consultation with doctors, protection of privacy, and public education and enlightenment on genetic diseases. The same principle should be applied to Japan.

Prevention of genetic diseases

The principle that no treatment is better than prevention can be applied to any disease. Next come early diagnosis and early treatment. In line with this principle, research is being made on safe and reliable fetal diagnostic tests at the earliest stage of pregnancy and some measures are being employed. In Japan, amniocentesis, chorionic villus sampling, fetoscopy, cordo-centesis and ultrasonography are used as diagnostic techniques. The Japan Society of Obstetrics and Gynecology, in 1987, provided its members with guidelines regarding requests for these examinations by couples, the meaning of the examination, safety and other matters.

In 1987 amniocentesis was employed for 3,022 cases at 84 institutions, and chorionic villus sampling for 140 cases at 10 institutions. These examinations detected 10.2% of abnormalities in the chromosome analysis group, 25% of congenital metabolic abnormalities in the biochemical analysis group, and 7.1% of abnormalities in the DNA analysis group. (These incidence rates are high because the tests were conducted on patients at high risk of genetic diseases.)

In Japan, compared with other countries, no genetic disease shows a particularly high incidence rate. The incidence of some of the congenital metabolic diseases can be reduced by palliative therapy after birth. However, there is no established therapy for most of the diseases. The only currently available way is elective abortion, in the light of the incidence rates of diseases and anticipated degree of impairment, and with consultation of doctors, based on requests of patients and their families.

Abortion in Japan

The actual incidence of elective abortion in Japan is not known; the incidence reported by the government is shown in Table 1. The incidence of abortion in other countries, including Eastern European countries, is shown in Table 2 for comparison. The actual number of elective abortions in Japan today, which is estimated to be double the reported number, almost equals the number of live births. The incidence of abortion for genetic reasons is presumed to be quite low compared with that of total elective abortion.

Table 1. Ratio of elective abortions to live births, 1955-88

Year	Elective Abortion (A)	Live Births (B)	A/B
1955	1,179,143	1,730,692	0.6761
1960	1,063,256	1,606,041	0.6620
1965	843,248	1,823,697	0.4624
1970	732,033	1,934,239	0.3785
1975	671.597	1,901,440	0.3532
1980	598,084	1,576,889	0.3793
1984	568,916	1,489,780	0.3819
1985	550,127	1,431,577	0.3843
1986	527,900	1,382,946	0.3817
1987	497,756	1,346,658	0.3696
1988	486,146	1,314,006	0.3670
Average	700,836	1,751,696	0.4362

Table 2. Ratio of abortion to 100,000 population in some countries

Year	Hungary	Czecho-slovakia	Bulgaria	Poland	Yugoslavia	Japan	G.D.R.	Sweden	Denmark
1955	35	2	4	0.5		131	0.73	6.3	12.3
1956	83	2.5	31	6.8		129	0.59	5.3	
1957	123	6	43.5	12.5		123	0.58	4.6	
1958	146	46	51.5	15	34	124	0.55	3.4	8.7
1959	152	60	59.5	27	50	119	0.55		
1960	163	65	72.2	50.5			0.5		
1961	170	69	88.5	48.5			0.45		

Here, I should like to discuss the history of abortion and birth control in Japan. The population changed very little in the middle hundred years of the Edo era; afterwards it increased slightly: in 1852, at the end of the Edo era, it showed only a 2.9% gain over 130 years. Besides the natural suppression of population growth by natural disasters and famines, frequent revolts by farmers, and the increase in deaths from various infectious diseases, including smallpox, abortion was a factor in suppressing population growth. However, also related was *Mabiki* or *Ko-Keashi,* which meant abortion or the killing of the newborn by a woman who did not belong to the family but who bathed the newborn. Poverty and congenital malformation of the baby are supposed to be the causes of these "executions". At that time, besides the woman who delivered the baby — the maternity woman — there was also the "washing woman", who corresponded to the midwife in modern times, and who was thought to play the role of regulator, distributing babies between the world of the live and the world of the dead, according to her judgement, in which she took the number and the quality of babies into account. However the maternity woman occasionally played the role of the "washing woman" at the same time. This custom continued into the end-stage of the Edo era, when contact with Western Christian civilization affected Japan, and the introduction of notification of births, as well as a decree of the Shogun Government, made the custom decline, which was one reason for a slight increase in population afterwards.

After World War II a rapid increase in population density occurred because of reduction of the territory and the return of Japanese nationals from abroad, together with an increase in the birth-rate associated with the increase in the numbers of retired service-men. Shortage of daily necessities and devastation of land were other factors. Against this background the Eugenic Protection Law was passed and took effect in 1949. Legal abortions began on the basis of Article 14-4 of the law, which allows a trained doctor to abort a fetus in cases in which the mother's health risked being harmed for economic reasons. This provision resulted in the failure of the law to prevent the crime of illegal abortion.

Since 1969, when Japan had restored its economic activity, a bill has three times been introduced in the Diet proposing the deletion of the economic reason for abortion and the addition of fatal disorder as a reason for elective abortion, from the standpoint of respecting the life of the fetus, but the bill was always defeated by the opposition of doctors and women. Still, the deep-rooted controversy between the right of the fetus to live and the mother's liberty and rights continues in Japan as well as in the United States.

The Eugenic Protection Law has a provision which allows an elective abortion if a pregnant woman or her relative has any of some 20 kinds of hereditary disorder, including Huntington's chorea, but has no so-called "fetal provision", which would allow elective abortion if a hereditary disorder of the fetus were diagnosed. Abortion seems to be carried out in this case under a broad interpretation of the provision about the economic reason, when the pregnant woman wishes it, although it is difficult to explain the logic of such interpretation.

The best way to prevent rationally a hereditary disorder is either to remain single or to adopt an adequate birth-control method. The only administrative policy which may be practised in the future is to promote the spread of knowledge of genetics, the enlightenment of the people, and proper genetic counselling before pregnancy.

Neonatal screening and genetic counselling

Technical genetic procedures and treatment of the fetus, which are radical methods of treating genetic disorders, are still at the stage of research, and this research needs more budgetary resources. The policy that should be practised for hereditary disorders is early diagnosis and treatment of the newborn. Therefore the national and local governments have instituted a mass screening programme for six inherited metabolic diseases, to be carried out at the time of registration of live births, for 15 years. As shown

Table 3. Result of screening of newborns in 1977-87 in Japan (from report of the Ministry of Health and Welfare)

Disease	Number examined	Number of patients found	Frequency
Phenylketonuria	13,541,228	179	(1/75,600)
Galactosemia	do.	222	(1/61,000)
Maple syrup urine disease	do.	32	(1/432,000)
Homocystinuria	do.	63	(1/214,900)
Histidinemia	do.	1,579	(1/ 8,600)
Cretinism	10,218,468	1,384	(1/ 7,400)

in Table 3, the incidence of these diseases in Japan is lower, except for histidinemia, than that in other countries. In addition to these six diseases, two other diseases are being screened for, and other diseases are being researched for screening. Most of the treatment of affected children is covered by the public insurance system. The essential conditions to be satisfied for neonatal screening are that: 1) the disease can be treated; 2) there is an established method of screening for it; and 3) it has a significant incidence. Another essential aspect to be taken into account by the government policy of screening is cost-effectiveness. The total cost of screening, and of diagnosis and treatment of the early discovered patients, amounts to about 32 billion yen, calculated on the basis of the yearly incidence rate, while in the case of delayed discovery of disease the total expenses would amount to about 130 billion yen. Thus, there is a saving of 98 billion yen and the cost-effectiveness ratio is about 1:4. From this point of view, a policy of neonatal mass screening should be further promoted.

Eugenic protection counselling services

Under Article 20 of the Eugenic Protection Law, the Ministry of Health and Welfare obliges every local government to set up a Eugenic Protection Counselling Center. The Genetic Counselling Center of the Japan Family Planning Association plays a core role in this service, which the Ministry of Health and Welfare subsidized to the amount of 12,000,000 yen in 1989.

The local governments have 836 Eugenic Protection Counselling Centers, of which 786 are attached to health centres. Besides, there are associated facilities, including university hospitals, which also give genetic counselling. The purpose of genetic counselling at these facilities is to promote the spread and the improvement of knowledge of genetics and other knowledge necessary for eugenic protection, and to promote the spread of, and guidance in, proper methods concerning regulation of conception. Concrete subjects of genetic counselling at the appropriate stage of pregnancy are the prevention of congenital disorders caused by environmental factors, counselling of hereditary high-risk groups, prenatal diagnostic procedures, neonatal examinations, and estimation of the risk for the next child, detection of the carrier state in infancy, treatment of the patient and family planning. Genetic counselling varies with regional peculiarities, family structure and degree of understanding of genetics and with philosophical, ethical, religious and other views.

Because counselling-physicians and public health nurses deal with very delicate problems, their attitudes are very important. For this reason the government directs counsellors to take special training in counselling. However, of 80 medical schools only two have departments of genetics and only one has a department of human genetics. Eight other schools have separate research institutes for genetics but no education in the subject. Hence, medical undergraduates and postgraduates have little opportunity to learn systematic human or clinical genetics. So far, genetic counselling has been practised by a small number of physicians who are interested in genetics, or by human geneticists. In 1977 the Ministry of Health and Welfare established a new project, the Family Planning Special Project, to provide genetic counselling services.

In 1989 some 76 hospitals or other facilities provided more than 800 centres where public health nurses could counsel. About 560 physicians received a nine-day training course, under government subsidy, to become genetic counselling doctors, and are already providing services; and 860 public health nurses, who have not received this training, are engaged in the service. The physician holds a high position from the patient's point of view and tends to direct the patient one-sidedly. The public health nurse is suitable for counselling on an equal relationship with the patient, and plays an important role in genetic counselling.

Future trends

More than 4,000 known hereditary disorders include various diseases, probably including hypertension, diabetes mellitus, cancers and others, thought to be related to physical constitution. These diseases are not directly brought about by the gene only, but by the interaction of the gene with the environment. This is a very important subject of research for the future. The radical treatment of genetic disorders needs artificial technical procedures involving DNA. Recently, a therapeutic procedure for adenosine-deaminase (ADA) deficiency has been developed in the United States, in which a harmless viral gene of mouse leukemia is to incorporate a gene producing the ADA enzyme and then made to infect the patient's leukocytes. The Biological Security Committee of the National Institutes of Health in the United States has recognized this procedure as a human experimental subject. This radical therapeutic procedure is expected to be applied to cancers and AIDS.

In Japan, however, a long period of discussion would be needed for the people to accept this method, as they have strong anxiety about, and resistance to, the artificial handling of genes. Seven years have already passed since the Diet began to consider the problem of brain death as a political subject in 1983, beginning with my interpellation there. More than 15 years have passed since the Japanese Association of Brain Waves established the criteria for judging brain death. Alone among developed countries, where brain death is accepted as proof of death and organs may be excised for transplantation, Japan does not legally recognize brain death as proof of death. It is a peculiarity of the Japanese way of thinking that in Japanese circumstances both brain death and "the fetal provision for abortion" are not recognized. I do not have time to explain this problem in detail, but briefly, the Japanese, who have generally rather scientific, analytic and rational views on non-living objects, are apt to be extremely emotional about life, especially human life and death, and adhere to tradition, without getting rid of it.

Also, they regard the subject of life and death as taboo, to be kept in a kind of sacred precinct, and avoid discussing death, keeping it far away as a disgusting thing. They tend to regard heredity as fate, with *a priori* dark senses, and prefer not to bring it into the light. Moreover, Buddhism and Confucianism do not have any literal behavioural norms, which Christianity and Islam have. The samurai were prepared to be killed but did not consider death itself. The Japanese seem to regard so-called *Wa*, or fellowship, as the best way to maintain the order of society, which prefers the acceptance of others' emotions with silence rather than discussion.

Under such a structure of deep consciousness, it is very difficult to apply a strict method of implementing a policy concerning life and death. The Japanese are a nation of most singular people. They are accustomed to a uniform style in both form and behaviour. They seem to have an intense sense of disharmony about form and behaviour that differ from the general way. Those with genetic disorders and their families find such a society difficult to live in. Therefore, they tend to hide the fact and keep it secret. They do not want a fetus to be aborted as one with a genetic disease. However scientific medicine advances in the future, it will be impossible to get rid of genetic disorders, as even a normal person has from five to eight abnormal genes. It should be natural therefore to assume that coexistence with the minority who are handicapped, the aged with disability and weak children is the ordinary natural condition. A Japanese policy on genetic disorders must take account of these aspects of Japanese consciousness.

124

Ackowledgements

I owe the preparation of this address to Professor T. Mori, Kyoto University School of Medicine; Associate Professor K. Suzumori and Professor Y. Wada, Nagoya Municipal University of Medicine; Professor K. Ohkura, Tokyo Medical and Dental College; Dr T. Ogata and Dr H. Endo, members of the Ministry of Health and Welfare, and members of the Division of Health of Mothers and Children.

GENE THERAPY FOR INBORN ERRORS OF METABOLISM

R. Parkman* and D.B. Kohn*

Bone marrow transplantation has an established role in the treatment of some genetic diseases[1]. The genetic diseases for which allogeneic bone marrow transplantation has had the greatest application are β-thalassemia major, an inborn error of hemoglobin synthesis, and inborn errors of metabolism, including adenosine deaminase (ADA) deficiency (severe combined immune deficiency), glucocerebrosidase deficiency (Gaucher's disease), α-iduronidase deficiency (Hurler syndrome) and iduronate sulphatase deficiency (Hunter syndrome). When histocompatible donors are available, patients, who receive cyclophosphamide to eliminate their lymphoid stem cells and busulfan to eliminate their hematopoietic stem cells, are uniformly engrafted with donor lymphoid and hematopoietic stem cells[2]. Analysis of bone marrow transplantation for children with thalassemia-β major reveals a greater than 90% disease-free survival; non-engraftment and graft-versus-host disease are the principal causes of failure[3]. Allogeneic bone marrow transplantation can reverse the non-central nervous system (CNS) symptomatology of patients with ADA deficiency, Gaucher's disease, Hurler syndrome and Hunter syndrome[4-6]. Controversy still exists, however, as to the impact of successful donor stem-cell engraftment on the CNS function of patients with inborn errors of metabolism[7]. The success of bone marrow transplantation in stabilizing the CNS manifestations of inborn errors of metabolism may depend on when transplantation occurs, since successful engraftment will not reverse pre-existing CNS damage. Longer follow-up of the patients already transplanted is needed before definitive statements can be made about the effects of bone marrow transplantation on the natural history of the CNS manifestations of inborn errors of metabolism.

Histocompatible bone marrow transplantation is limited to the 20-25% of patients with a histocompatible sibling or phenotypically identical parent. Thus, the majority of affected children cannot benefit from bone marrow transplantation. Therefore, many investigators have entertained the possibility of inserting the normal copy of the defective gene into autologous bone-marrow cells, which could then be returned to the patient[8,9]. The successful use of genetically corrected autologous bone marrow would mean that all affected patients could be cured. The use of autologous bone marrow would eliminate the problems associated with graft-versus-host disease, which is still the major post-transplant related problem.

* Division of Research Immunology/Bone Marrow Transplantation, Children's Hospital, Los Angeles and Department of Pediatrics and Microbiology, University of Southern California School of Medicine, Los Angeles, CA, U.S.A.

Histocompatible allogeneic bone marrow transplantation requires the ablation of both the patient's lymphoid and hematopoietic stem cells with concomitant myelosuppression and immunodeficiency. If genetically-corrected autologous bone marrow were used, it might be possible to selectively ablate the lymphoid or hematopoietic stem cells, depending upon the disease treated. Thus, the treatment of ADA deficiency would require the ablation of only the lymphoid stem-cells, while treatment of thalassemia would require the ablation of only the hematopoietic stem-cells.

ADA deficiency

Two model diseases of gene therapy for inborn errors of metabolism are ADA deficiency, a model for disorders of the lymphoid system, and Gaucher's disease, a model for disorders of the hematopoietic system. Allogeneic bone marrow transplantation is curative therapy for children with severe combined immune deficiency due to ADA deficiency[4]. Following transplantation, the patient's T-lymphocytes are capable of normal immunological function, even though the other recipient cells, including hematopoietically-derived cells (monocytes, erythrocytes, granulocytes) continue to be ADA-deficient. These clinical observations indicate that it is the intracellular content of ADA that determines whether normal T-lymphocyte differentiation and function are possible.

For those patients with ADA deficiency who do not have histocompatible donors, alternative therapies include T-lymphocyte-depleted, haploidentical bone marrow transplant and bovine adenosine deaminase coupled to polyethylene glycol (PEG-ADA). In stable patients with ADA deficiency, T-lymphocyte depleted, haploidentical transplants with parental bone marrow have a success rate of 67%, a figure similar to that seen with histocompatible transplantation[10]. However, some patients have established opportunistic infections, particularly with viral pathogens, making the cytoreduction that is required prior to T-lymphocyte-depleted transplantation clinically difficult.

ADA deficiency is a heterogeneous disorder with varying levels of endogenous ADA activity (0.1 to 5% of normal), in part due to the heterogeneity of the primary defect[11]. The presence of low levels of ADA can result in some immunological function, inadequate for normal production against infectious organisms. The initial attempt to provide an exogenous source of ADA consisted of the transfusion of normal ADA-containing erythrocytes. The transfusions normalized the plasma levels of deoxyadenosine metabolites, but resulted in no antigen-specific immune function. The long-term administration of erythrocytes, however, causes significant clinical problems. Patients are now being treated with PEG-ADA as a source of exogenous ADA. The administration of PEG-ADA has also resulted in the normalization of plasma deoxyadenosine metabolites[12]. Some patients have had clinical improvement, with increased weight gain and decreased infections. *In vitro* studies have shown an improvement in mitogen-induced T-lymphocyte blastogenesis. However, patients have not

127

routinely developed antigen-specific immune function as measured *in vitro* by antigen-specific T-lymphocyte blastogenesis or *in vivo* by the production of specific antibodies. Thus, patients who are not eligible for T-lymphocyte-depleted, haploidentical transplants and who have not developed antigen-specific immune function on PEG-ADA require alternative forms of therapy.

ADA-deficient T-lymphocytes have normal ADA levels after the retrovirally-mediated insertion of the normal ADA gene[13]. The ADA functions normally, as demonstrated by the fact that the sensitivity of ADA-deficient T-lymphocyte cell-lines to 2' deoxyadenosine is normalized after the insertion of the normal ADA gene. In contrast to erythrocytes, monocytes and granulocytes, all of which have relatively short *in vivo* half-lives, some T-lymphocytes are long-lived (more than 20 years). The insertion of a normal ADA gene into the peripheral blood T-lymphocytes of ADA-deficient patients and their infusion into the patient might, therefore, result in the persistence of the "corrected" T-lymphocytes in the patient for significant periods of time. One potential mechanism by which the ADA-containing T-lymphocytes might produce immunological improvement is that the presence of normal ADA levels within a nucleated cell would provide more effective detoxification of the deoxyadenosine metabolites than circulating extracellular PEG-ADA or ADA within non-nucleated cells (erythrocytes), resulting in increased *in vivo* survival of the ADA-containing T-lymphocytes. A second hypothesis is that ADA-deficient T-lymphocytes, after the insertion of a normal ADA gene, are capable of antigen-specific T-lymphocyte function. *In vivo* evidence in SCID mice has demonstrated that the peripheral-blood T-lymphocytes of an ADA-deficient patient, into which a normal ADA gene had been inserted, were capable *in vivo* of producing normal human immunoglobulin. Further, the transplanted T-lymphocytes persisted *in vivo* for a month, and alloreactive and antigen-specific T-lymphocytes could be detected[14]. In contrast, when non-transfected ADA-deficient peripheral-blood T-lymphocytes were tranplanted, no human immunoglobulin was detected, and no alloreactive or antigen-specific T-lymphocytes were present. These *in vivo* murine experiments indicate that the presence of a normal ADA gene may permit immunological function and prolonged *in vivo* survival by the ADA-deficient T-lymphocytes, and are the basis of a proposed clinical trial, in which peripheral-blood T-lymphocytes from ADA-deficient patients will be stimulated *in vitro* with monoclonal antibodies to CD3 and then transfected with a retroviral vector containing a normal ADA gene. The activated T-lymphocytes will then be expanded *in vitro* with IL-2 until adequate cell numbers are obtained and then infused into the patient. Because the ADA-containing vector also contains the neomycin-resistant gene, it is possible to select the *in vitro*-expanded T-lymphocytes for only those cells containing the normal ADA and neomycin genes by the addition of G418. Patients can then be analysed for the persistence of circulating T-lymphocytes containing the human ADA gene and improvement of their T-lymphocyte function, including clinical improvement (weight gain, cessation of established viral infections) and development of antigen-specific immune function (antigen-specific T-lymphocyte blastogenesis, production of specific antibody). Such a protocol has been proposed by investigators at the National Institutes of Health.

The insertion of the normal ADA gene into peripheral-blood T-lymphocytes eliminates the need for ablating the patient's hematopoietic and lymphoid stem-cells. Thus, the risk of the proposed gene-therapy trial is minimal since the patients will receive no chemotherapy and will be monitored only for the function of the infused T-lymphocytes. The potential efficacy of the administration of ADA-containing T-lymphocytes is possible only because some T-lymphocytes are long-lived. If T-lymphocytes, like granulocytes, monocytes and erythrocytes, had a relatively short *in vivo* survival, the transfection of the normal ADA gene into mature T-lymphocytes would have no potential clinical advantage.

Gaucher's disease

In contrast to the use of gene therapy to treat the immune deficiency associated with ADA deficiency, the correction of inborn errors of metabolism affecting the hematopoietic system (Gaucher's disease, Hurler syndrome, Hunter syndrome) by the insertion of the normal gene into circulating hematopoietically-derived elements is not possible, because of the short half-life of the mature hematopoietic cells. In Gaucher's disease, the absence of functional glucocerebrosidase results in the accumulation of non-degraded membrane lipoproteins in the cells of the reticuloendothelial system. Thus, even if it were possible to insert the normal glucocerebrosidase gene into circulating monocytes, their short *in vivo* survival would not result in any clinical benefit. Histocompatible bone-marrow transplantation has demonstrated that the turnover of tissue macrophages (Kupffer cells, alveolar macrophages, microglial cells) takes four to six months[15]. Therefore, the only way to have persistence of hematopoietic cells, into which the normal glucocerebrosidase gene has been inserted, is to replace the patient's abnormal hematopoietic stem-cells with pluripotent hematopoietic stem-cells into which the normal glucocerebrosidase gene has been inserted. An *in vivo* model for the detection and quantification of pluripotent human hematopoietic stem cells does not yet exist. Investigators using human hematopoietic stem-cells must limit themselves to long-term bone-marrow cultures. If *in vivo* analysis is required, it is necessary to study species other than man (mice, monkeys)[16]. Investigators studying Gaucher's disease have transfected the human glucocerebrosidase gene into the pluripotent murine hematopoietic stem-cells[17]. The only clear demonstration that a gene has been transfected into a pluripotent hematopoietic stem-cell is the sequential transplantation of bone marrow to demonstrate that the bone-marrow cells into which the gene had been transplanted in the first recipient are capable of repopulating a second recipient with multi-lineage progenitors. Investigations have demonstrated that secondary recipients of bone marrow containing the human glucocerebrosidase gene express the transfected human gene. Long-term human cultures of normal and Gaucher's bone marrow have shown the continued expression of the transfected normal glucocerebrosidase gene by both Northern analysis and immunoblotting. The maximum infection of hematopoietic progenitors occurs when bone marrow cells are stimulated *in vitro* with hematopoietic growth factors, particularly the combination of IL-3 and IL-6, which can increase the percentage of pluripotent hematopoietic stem-cells in cycle[18]. These *in vivo* and *in vitro* experiments with the glucocerebrosidase gene indicate that Gaucher's disease may be the first inborn error of the hematopoietic system in which gene therapy will be attempted.

Conclusion

Gene therapy for inborn errors of metabolism is not yet a clinical reality. Ultimately, gene therapy will require the insertion of the normal gene into pluripotent bone-marrow stem-cells or other stem-cell populations that have the capacity for self-renewal. In the interim, the insertion of normal genes into mature cells with long *in vivo* half-lives may be of clinical benefit. The proposed studies for the insertion of the normal human ADA gene into ADA-deficient peripheral-blood T-lymphocytes would be such a trial.

References

1. Parkman, R. The application of bone marrow transplantation for the treatment of genetic diseases. *Science* 1986; **232**: 1373-8.
2. Parkman, R. *et al*. Busulfan and total body irradiation as antihematopoietic stem cell agents in the preparation of patients with congenital bone marrow disorders for allogeneic bone marrow transplantation. *Blood* 1984; **64**: 852-7.
3. Lucarelli, G. *et al*. Bone marrow transplantation in patients with thalassemia. *N. Engl. J. Med*. 1990; **322**: 417-21.
4. Parkman, R. *et al*. Severe combined immunodeficiency and adenosine deaminase deficiency. *N. Engl. J. Med*. 1975; **292**: 714-9.
5. Rappeport, J.M. & Ginns, E.I. Bone marrow transplantation in severe Gaucher's disease. *N. Engl. J. Med*. 1984; **311**: 84-8.
6. Hobbs, J.R. *et al*. Reversal of clinical features of Hurler's disease and biochemical improvement after treatment by bone-marrow transplantation. *Lancet* 1981; **2**: 709-12.
7. Hugh-Jones, K. Psychomotor development of children with mucopolysaccharidosis Type 1-H following bone marrow transplantation. In Krivit, W. and Paul, N.W. (eds): *Bone Marrow Transplantation for Treatment of Lysosomal Storage Diseases*, Vol. 22. New York, Alan R. Liss, 1986; pp. 25-29.
8. Anderson, W.F. Prospects of human gene therapy. *Science* 1984; **226**: 401-9.
9. Kohn, D.B. *et al*. Gene therapy for genetic diseases. *Cancer Invest*. 1989; **7**: 179-92.
10. O'Reilly, R.J. *et al*. The use of HLA-non-identical T-cell-depleted marrow transplants for correction of severe combined immunodeficiency disease. *Immunodeficiency Rev*. 1989; **1**: 273-309.
11. Markert, M.L. *et al*. Adenosine deaminase and purine nucleoside phosphorylase deficiencies: Evaluation of therapeutic interventions in eight patients. *J. Clin. Immunol*. 1987; **7**: 389-99.

12. Hershfield, M.S. *et al*. Treatment of adenosine deaminase deficiency with polyethylene glycol-modified adenosine deaminase. *N. Engl. J. Med.* 1987; **316**: 589-96.
13. Kantoff, P.W. *et al*. Correction of adenosine deaminase deficiency in cultured human T and B cells by retrovirus-mediated gene transfer. *Proc. Natl. Acad. Sci. U.S.A.* 1986; **83**: 6563-7.
14. Bordignon, C. *et al*. In vivo and in vitro models for analysis of retroviral-mediated expression human ADA in ADA deficient cells from patients affected by severe combined immunodeficiency. *J. Cellular Biochem.* 1990; Suppl. **14A**: 366.
15. Gale, R.P. *et al*. Bone marrow origin of hepatic macrophages (Kupffer cells) in humans. *Science* 1978; **201**: 937-8.
16. Kantoff, P.W. *et al*. Expression of human adenosine deaminase in non-human primates after retroviral mediated gene transfer. *J. Exp. Med.* 1987; **166**: 219-34.
17. Weinthal, J. *et al*. Expression of human glucocerebrosidase in hematopoietic cells by retroviral vectors. *Pediatr. Res.* 1990; **27**: 137A.
18. Nolta, J.A. & Kohn, D.B. Comparison of the effects of growth factors on retroviral vector-mediated gene transfer and the proliferative status of human hematopoietic progenitor cells. *Human Gene Therapy*. In press.

SOME ETHICAL IMPLICATIONS
OF HUMAN GENE THERAPY

Theodore Friedmann[*]

It is well recognized that scientific and technological advances may often lead to serious dilemmas in the use of the expanded knowledge and in the delivery of improved technology to human beings, for both benevolent and malevolent purposes. One of the most obvious examples of the development of moral, ethical and public-policy problems driven by unanticipated or uncontrolled directions in technology was the sudden appearance in the mid-20th century of the human capability to manipulate the energy of the atom. This power has spawned a long list of potential benefits and abuses, some of which have become reality, while others remain theoretical. All have led to extensive ethical and policy debate.

Amazing technological advances in the biomedical sciences over the past several decades have also produced areas in which human choices between conflicting needs and expectations — ethical and public-policy choices — must be made. These advances span the entire spectrum of modern biomedicine; they include not only those associated with universal availability of basic medical care but also those associated with the delivery of highly technological services, including the use and abuse of life-support systems, organ transplantation, anti-reproductive technologies and many others.

For many reasons, genetic technology has been prominent in the debates over the application of science to human problems. From the Biblical injunctions on the choice of a wife to the most modern manipulations promised or, as some would have it, threatened by modern molecular biology, genetics among all the sciences has continued to hold a place of special interest and concern — and not without justification. After the discovery of the laws of inheritance by Mendel in 1865, the potential application of genetic science to human affairs took a number of paths, some of which obviously were to lead to expanded knowledge and improvements of the human condition, but others of which were destined to lead to troublesome and ethically ominous developments. The birth of "scientific" underpinnings to the concepts of eugenics in England during the last quarter of the 19th century and in the United States during the early part of the 20th century[1] was to lead in those two countries not only to peculiar concepts of the heritability of such human traits as laziness, a love of the sea and hopelessness, but also to the incorrect demonstration of many other genetic "differences" among humans of different ethnic, social and intellectual groups. The inevitable result was that many malevolent social program-

[*] Center for Molecular Genetics and Department of Pediatrics, and Whitehill Chair of Biomedical Ethics, University of California at San Diego School of Medicine, La Jolla, California, U.S.A.

mes, including discriminatory immigration policies and compulsory sterilization laws, were supported and enacted. The Supreme Court of the United States, in the Buck vs. Bell decision of 1927, upheld one such compulsory sterilization law with the ringing phrase, "Three generations of imbeciles are enough." The abuse of so-called genetic information reached a zenith in the genocidal programmes of Nazi Germany, and the role of the German geneticists, psychiatrists and anthropologists has been thoroughly examined and illuminated[2].

At an apparently increasing and unremitting pace, modern medicine, genetics and molecular biology are continuing to provide a treasure of new techniques that will permit an unprecedented understanding of genetic mechanisms responsible for many human diseases and many human traits, and the early detection of disease-related genetic defects and factors that contribute to disease susceptibility. Many such tools have already been used for the improved well-being of many. But, as with all technology, some have been used in premature and ill-conceived programmes, as in the case of the mis-steps inherent in the efforts in the United States programmes to detect the XYY phenotype and in the sickle-cell detection programmes of the 1970s[3]. The stimulus provided by the international effort now being mounted to characterize the human genome will almost certainly increase further the speed with which science will develop and make available the tools for improved understanding and even treatment of human disease, and even for an understanding of the genetic nature of the human creature[4].

Gene therapy

One of the areas in which modern molecular genetics seems to hold the greatest medical promise is the development of conceptually new approaches to the treatment of human diseases and disorders. From its very earliest beginnings to the present day, medicine has been forced to be content always with treating the consequences of genetic defects rather than with a correction of the underlying defect itself — the mutant gene. This limitation is soon to be a thing of the past, thanks to the development of the concepts and techniques for correcting genetic defects themselves — gene therapy[5]. Beginning in the early 1970s, scientists first proposed and now have made feasible a number of methods to introduce normal genetic material into defective cells for therapeutic purposes[6].

Model studies with a number of model systems have suggested strongly that such foreign genes can function appropriately in their new host-cells to correct genetic deficiencies permanently and stably. If this becomes feasible, genetic treatment of many disorders, including not only the relatively rare inborn errors of metabolism but also such major and common diseases as cancer, atherosclerosis, Parkinson's, Alzheimer's and other neuropsychiatric diseases, will offer major advantages over conventional therapies. This would seem to be an obvious and unqualified good. Yet, even as the first human clinical trials are under way, the field of human gene therapy

133

has not been without controversy. Why have programmes for human gene-therapy come under such careful scrutiny and engendered such widespread debate? What are the ethical questions and dilemmas posed by this new approach to human disease?

Recent mis-steps

The most immediate and obvious explanation for unease at the prospects of human gene therapy is that, already in its very brief history, the field has seen two very controversial studies that have been heavily criticized not only on technical grounds, but also and even more heavily on ethical grounds. The first involved the deliberate infection of two severely ill children suffering from hyperargininemia with the Shope papilloma virus[7], a treatment based on rather inadequate evidence that the virus was capable of introducing a foreign arginase gene into cells to replace the absent arginase function responsible for the devastating disease. The second study involved the injection of cloned globin genes into several patients suffering from thalassemia, a very severe disorder in which the patients are unable to produce the oxygen-carrying blood-protein hemoglobin because of mutations in their own globin genes. The physician-scientist who carried out these studies was severely reprimanded by the National Institutes of Health, lost much of his federal funding and had to step down from his administrative post at his university[8,9]. As a direct result of this incident, the National Institutes of Health now requires that all genetic therapeutic studies with human patients first be reviewed and approved by the Recombinant DNA Advisory Committee of the National Institutes of Health, as well as by a number of other university and federal review bodies.

Important as these two incidents were, probably a more important factor underlying our rather general unease with genetic manipulation of humans lies in the threats that genetic information seems to pose at the religious and metaphysical levels. Critics have claimed that the approach constitutes an impermissible and arrogant tampering by man with nature and with its slow, deliberate and proven mechanisms for genetic evolution and selection. The human genome, they have argued, ought to be inviolable and ought not to be tampered with, even for therapeutic purposes. They have argued that we know too little of all the functions of our genetic information to be able to predict the consequences of deliberate change. The resulting current ethical debate regarding gene therapy in humans revolves around two major questions: the question of somatic vs. germ-line modification, and the question of the identification of the human traits or disorders that constitute suitable targets for genetic modification.

Somatic-cell vs. germ-cell modification

The first question is whether designed genetic change should involve only the somatic cells of patients, or also the germ cells. In other words, should gene therapy be restricted to only a given individual and his cells or should it be used also to modify the genes in the reproductive cells to ensure that the genetic correction will be passed to the patient's progeny and therefore to future generations. The latter has the theoretical advantage of eliminating the appearance of a genetic disease in many generations instead of in a single patient, and in the event that we might all agree that a particular disease-related gene is providing no subtle and unrecognized but important evolutionary or physiological function, it does seem an effective method of disease control.

At first the answer to the question of germ-line manipulation may have seemed rather simple, since it was based on the likelihood that techniques for stable, safe and efficient introduction of genes into germ cells would probably not readily become available. However, this now seems rather unlikely, and it therefore seems very unlikely that the answer to the ethical question of the propriety of human germ-cell modification will be only a technical one for long. There are a number of well-studied animal models in which foreign genetic information can be introduced into the germ cells in a way that allows them to be passed on, apparently in stable and expressed form, in the germ line to later generations. Admittedly, the methods are not yet highly efficient in these animal studies, but it does seem very likely that the obstacles to similar phenomena succeeding in humans, while formidable, will prove to be merely technical and that the techniques developed for human cells will not be so very different from those that already exist for lower forms of mammalian life.

A large number of religious, ethical, public policy and scientific bodies have examined and discussed the feasibility of somatic- and germ-cell therapy[5]. The vast majority have concluded that the genetic modification of somatic cells for therapeutic purposes raises very few, if any, serious or novel ethical concerns not already raised by many other forms of more conventional medical intervention. However, most of these same groups have not endorsed germ-cell modification in humans as an appropriate procedure in human beings, on both technical and ethical grounds. Rather, they have concluded that, so long as the results of the procedure cannot be known with absolute certainty for each and every human who is to result from genetically modified germ-cells, the danger of producing very much more harm than good should preclude any attempts to carry out such manipulations in humans. They point to the availability of alternative theoretical approaches, such as the selection and re-implantation of only proven normal human embryos produced by *in vitro* fertilization. One major flaw in this argument, of course, is that such very high-technology medicine would not be available equitably to vast numbers of potential patients. A simple, once-and-for-all correction of a genetic disorder would certainly be more efficient and, in some cases, probably even desirable, especially for disorders in which damage is early and irreversible, and otherwise untreatable.

Some argue that on ethical grounds we cannot allow such manipulations, even if the technical and safety questions were easily solved. They maintain that it would be unacceptable ethically for us of this generation and this time to make reproductive and evolutionary decisions for future generations, to deprive future human beings of the autonomy of making their own choices. Of course, we make such decisions constantly in all realms of human work, albeit through different mechanisms and with different implications for the speed with which the effects will become evident. We make environmental, political, social, economic and even medical decisions with very serious implications for future generations. We cannot do otherwise, and it would be inappropriate for us to abrogate that responsibility - failure to act in the face of a need or in the cause of prevention of misery and suffering is every bit as much an ethical choice as acting appropriately or inappropriately, wisely or foolishly. Of course, the decisions we make must be decisions that protect not only the autonomy of future generations, but their well-being as well.

It would be unwise and certainly premature for us, individually or collectively through international pronouncements, to foreclose the possibilities for germ-cell genetic correction in the future. However, we can all agree that, if germ-line correction of some kinds of human disease is to find a place in future therapies, very much more research must be done and much more must be understood about human reproductive mechanisms before any attempts can be made at the deliberate genetic modification of human germ-cells.

What traits are targets for genetic modification?

In answer to the second question, there is little doubt that gene therapy will become available and useful for some kinds of genetic diseases, and will eventually come to be used. There is little disagreement that most model diseases now being imagined as potential targets for gene therapy represent appropriate target diseases, since they are conditions that most of us, acting as moral agents, agree represent true "disease" — "errors" in the flow of genetic information. Medicine as a human activity in most Western society is built on the theological, moral, ethical and public-policy foundation that disease and its human suffering are to be fought and conquered. We readily accept the use of other, now "traditional" therapeutic tools — scalpels, antibiotics and our vast collection of medicines and drugs. We shall soon come to accept genes themselves in exactly the same way.

However, the very same methods that will allow us to manipulate disease may also enable us to modify other traits that may not represent such obvious examples of "disease", and which not all moral agents can agree should be attacked at all, let alone at the genetic level. We are then faced with the problem of making the sometimes very difficult distinction between therapy and what might be called eugenic genetic manipulation — the deliberate and designed enhancement or suppression of human non-disease traits — intelligence, growth properties, strength, aspects of personality. For instance, does an intelligence quotient of 90 reflect a disorder calling for correction? Or of 75? Or of 50? At what point does this variable human quantity change into a defect to be treated? The answer to these

136

and many similar questions is further complicated by the fact that even our available therapies are not restricted to treatment of overt disease. Through the use of such established forms of medical treatment as surgery and drug therapy for a wide variety of conditions other than disease, our societies already accept the concept of cosmetic or supportive treatment — that is, the notion that well-being and health are not always so clearly distinguished from the absence of obvious disease. Nevertheless, many commentators have argued that gene therapy is sufficiently different from other therapies that it should be reserved for the treatment of frank disease. I think that there is much merit in this argument, and there is no doubt that, until techniques are thoroughly understood and tested, it will and should prevail. Gene therapy will be used for serious disorders with no equally effective alternative therapies; such disorders include cancer, metabolic diseases, Parkinson's disease and other neuropsychiatric disorders. It is equally sure that, eventually, but not yet, genetic tools will be used for many of the conditions that call now for other kinds of treatment.

Again, let us not paint ourselves into a conceptual corner by stating that some human traits, such as intelligence and memory and other quantitative human characteristics, that might be amenable to genetic manipulation ought not be appropriate conditions for genetic modification even though they may not represent obvious disease states. The time for such kinds of genetic manipulation is obviously much farther off than that for gene therapy for classical genetic disease, and we may find that they will indeed not become feasible or effective. We may therefore even choose not to use genetic tools in this way. However, as in the case of germ-cell modification, our minds and imaginations should remain open, and we should be prepared to use the powers and tools of molecular genetics for the amelioration of human disease and the relief of suffering. We should ensure that we come to use the powers and tools wisely, compassionately and for the enhancement of human freedom and health.

Conclusion

The tools of molecular genetics are poised for an attack on human disease with an efficiency unprecedented in the history of medicine. We are being handed techniques that will make it possible for the first time to bring about truly curative approaches to many human ailments through a direct attack on the root cause of a disorder — a mutant gene. The goal of these concepts is therapy and not the design of genetically ideal human beings. To belittle the need for definitive genetic approaches to disease management is to deny the most rational of all treatment principles — to correct what is defective. Gene therapy will come because it is needed, and because in the long run it offers the best hope for the amelioration of very much of human suffering. Let us take advantage of its strengths but not lose sight of the potential for problems such as those associated with the questions of who is the patient and what is the disease.

References

1. Kevles, D.J. *In the name of eugenics*. Alfred A. Knopf, New York, 1985.
2. Muller-Hill, B. *Murderous science*. Oxford University Press, Oxford, 1988.
3. Holtzman, N.A. *Proceed with caution: predicting genetic risks in the recombinant DNA era*. Johns Hopkins University Press, Baltimore, 1989.
4. Friedmann, T. The human genome project — Some implications of extensive "reverse genetic" medicine. *Am J Med Genet* 1990; **46**: 407-14
5. Friedmann, T. Progress toward human gene therapy. *Science 1989;* **244**: 1275-81.
6. Friedmann, T. and Roblin, R. Gene therapy for human genetic disease? *Science* 1972; **175**: 949-55.
7. Rogers, S. Gene therapy: A potentially invaluable aid to medicine and mankind. *Res. Comm. Chem. Path. Pharm* 1971; **2**: 587-99.
8. Anderson, W.F. Gene therapy. *JAMA* 1981; **246**: 2737-39.
9. Wade, N. UCLA gene therapy racked by friendly fire. *Science* 1980; **210**: 509-11.

A GENE TRANSFER EXPERIMENT

Kunio Yagi*

Hereditary metabolic disorders due to lack of enzymes should become curable by expression of the appropiate genes transferred into the proper cells of patients. Also, the expression of genes directing the synthesis of certain biogenic substances might be useful for the cure of certain diseases. Such gene therapy consists of: (a) construction of a plasmid containing the desired gene, and (b) transfer of the plasmid into the proper cells.

When foreign genes are artificially transferred into cells, some recipients show transient or stable expression of these genes, resulting in the appearance of altered characteristics of cellular function. A wide variety of methods of gene transfer include the use of calcium phosphate, DEAE-dextran, microinjection, electroporation, retroviruses, protoplast fusion, cationic lipid and liposomes. Among these the use of liposomes seems to be the most preferable for *in vivo* experiments, since they have little or no toxicity. Liposomal vesicles with a large aqueous volume made by the reverse-phase evaporation method devised by Szoka and Papahadjopoulos[1] can encapsulate genes and seem to be the most useful tool for clinical application. As a prelude to gene therapy, we examined this method and found that some amount of DNA was damaged during the preparation and that the damage was due mainly to the sonication involved in the procedure. Therefore, we improved the preparation procedure so as to obtain less altered DNA entrapped in liposomes[2].

In addition, we devoted our efforts to obtaining liposomes better able to entrap and transfect DNA, since the efficiency of gene transfer by the use of liposomes is rather low without certain aids, such as glycerol shock, polyethylene glycol treatment or the addition of viral fusion proteins. Considering that both DNA and the cell surface possess negative charges, we considered that positively charged liposomes should be suitable for both efficient entrapment of DNA and the efficient transfection of DNA into cells. Thus we screened liposomes containing various positively charged lipids for the expression of the bacterial β-galactosidase gene[3].

At first, the integrity of liposome-entrapped DNA was studied. By the original reverse-phase evaporation method, we prepared liposome-entrapped pCH110 plasmids. Then the plasmid DNA was extracted from liposomes and analysed by agarose gel electrophoresis. The migration pattern of the plasmid DNA is markedly different from that of the plasmid before the entrapment in the liposomes. The plasmid before the entrapment was composed of a rapidly moving major component, assigned to the supercoiled

* Director, Institute of Applied Biochemistry, Yagi Memorial Park, Mitake, Gifu, Japan.

form, and a slowly moving minor component assigned to the relaxed form. The plasmid entrapped in liposomes had two components: a major component, apparently identical with the relaxed form, and a minor component identical with the linear form, of pCH110. This result can be interpreted as indicating that the conformation of the original plasmid had been altered by its entrapment in the liposomes, with the result that some of it was changed to its linear form. This alteration is assumed to be due to the sonication involved in the original method, since it is well known that sonication produces nicks in DNA strands.

Accordingly we tried to prepare liposome-entrapped plasmids without sonication of DNA. First, we prepared reverse-phase micelles by sonication, and then mixed them with the plasmids to form liposome-entrapped plasmids. The mean diameter of the liposomes was about 100 nm. The plasmid thus entrapped in liposomes was analysed in the same way as the plasmid entrapped in liposomes prepared by the original method. The migration pattern of the plasmids upon agarose gel electrophoresis, compared with the pattern of the plasmids before entrapment, indicated the major portion to be the relaxed form, and the minor portion the supercoiled form. The linear form found for the plasmids entrapped in liposomes prepared by the original method was not detected when the modified method was employed. The conformational change of pCH110 from the supercoiled to the relaxed form might have been induced by the extraction from the liposomes, since it is known that freezing and thawing also produce nicks in DNA strands to some extent, resulting in the production of the relaxed form with nicks. These results indicate that DNA entrapped in liposomes by the present method is, at least partly, undenatured.

Then the expression of the β-galactosidase gene encoded in plasmid pCH110 was studied. When the gene was transferred with liposomes prepared by the improved reverse-phase evaporation method, its expression was detected, but this was not the case with liposomes prepared by the original method. However, the transfection efficiency of liposomes prepared by the present method was still insufficient. Accordingly, we sought to construct more suitable liposomes that would be efficient in entrapment of DNA and in transfer of DNA into cells.

Since positive charges on the surface of vesicles seemed to be advantageous for the transfer of genes into the cells, we prepared various liposomes containing the positively charged lipids by the improved reverse-phase evaporation method and found that liposomes containing positively charged lipid N- (δ-trimethylammonioacetyl)-didodecyl-D-glutamate chloride (TMAG) gave highly efficient transfection. After repeated preliminary experiments, we found that TMAG/dilauroyl phosphatidyl-choline (DLPC)/dioleoyl phosphatidylethanolamine (DOPE) in a molar ratio of 1:2:2 yielded the most efficient expression.

Using the above liposomes, we examined the relationship between the amount of the plasmid added and that of the enzyme expressed. For this puropose, different volumes of a suspension of liposome-entrapped plasmids

were added to the medium covering the cells. After the incubation, expressed β-galactosidase was measured for its activity. The amount of the enzyme
expressed increased with increasing amount of the plasmid, but a further
increase in DNA concentration resulted in a rather large decrease in the
enzymatic activity. This fact may be ascribed to a toxic effect of the increased amount of lipids at higher plasmid levels.

In the light of the above results, we performed the transfection experiment, using a suspension of liposomes containing a suitable amount of
DNA. TMAG-containing liposomes, TMAG/DLPC/DOPE in a molar ratio
of 1:2:2, revealed the highest transfection efficiency. This efficiency ex-
ceeds that of the calcium phosphate precipitation method performed with
glycerol shock. The transfection efficiency of the liposomes containing
TMAG was affected by other components. When the DOPE in TMAG/
DLPC/DOPE-liposomes was replaced with dioleoyl phosphatidylcholine
(DOPC), a six-fold drop in the expression level was found; whereas with
dilauroyl phosphatidylethanolamine (DLPE) or dimyristoyl phosphatidylethanolamine (DMPE), a considerable expression level was still retained.
Further, if dimyristoyl phosphatidylcholine (DMPC) was employed instead
of DLPC, a 12-fold drop in potency occurred. The transfection efficiency
in the case of liposomes prepared without TMAG dropped enormously.
This drop was not due to the toxicity of lipids.

These results indicate the importance of the electrostatic interaction between the negatively charged cell surface and the positively charged surface
of the liposomal membrane for the transfection. In addition, the fact that
the various TMAG-containing liposomes showed different transfection ef-
ficiencies suggests that components of the liposomes other than TMAG
also play an important role in the transfection process, involving the incorporation of genes into the cells across the cellular membrane and transport
of genes to the proper site of the cells. To explain these differences, we
must examine the fate of the present liposomes during the transfection
process. This awaits further investigation.

As a control, we conducted the same experiment with empty liposome-
DNA complexes at the same ratio of amounts of DNA and lipids as that
used for the liposome-entrapped DNA. Empty liposome-DNA complexes
showed a definite transfection, and this can be ascribed to the similar effect
as observed in lipofection. However, the transfection efficiency of the complexes was lower than that of liposome-entrapped DNA. This indicates
that the entrapment of DNA by the present liposomes is due not only to
complex formation but also to encapsulation.

As to DNA-entrapment efficiency, the liposomes containing TMAG
revealed a 5- to 70-fold higher efficiency than those without TMAG. Taking into account the strongly positive charge of TMAG, the high efficiency
of the liposomes containing TMAG in terms of DNA entrapment can be
ascribed mainly to the electrostatic interaction between DNA and TMAG,
as is the case for lipofection. It should be noted, however, that the entrapment efficiency did not parallel the expression efficiency.

All the data reported in the present paper indicate that the transfection and expression of DNA can be achieved by simple addition of positively charged liposome-entrapped plasmids to the medium covering the cells without any additional manipulations such as glycerol shock. The expression level obtained by the present method was higher than that by the calcium phosphate method, especially in the case of a small amount of DNA. Another advantage of the liposomal vehicle is that a specific antibody or a ligand for a specific receptor can be attached to such liposomes in order to deliver the gene to the specific cell, organ or tissue desired. As an example, we entrapped plasmids containing the human interferon-β gene in such liposomes. Human glioma cells transfected *in vitro* with these liposomes produced human interferon-β. When the liposomes were coupled to a monoclonal antibody against glioma-associated antigen, the production of human interferon-β was increased more than tenfold. Then we examined the *in vivo* expression of the human interferon-β gene, using transplants of human glioma cells in nude mice, and found that the cells continuously produced high levels of human interferon-β cytostatic to themselves. All these methods presented here should have a potential utility in gene therapy.

References

1. Szoka, F. Jr. & Papahadjopoulos, D. Procedure for preparation of liposomes with large internal aqueous space and high capture by reverse-phase evaporation. *Proc. Natl. Acad. Sci. USA*. 1978; 75: 4194-8.
2. Haga, N. & Yagi, K. An improved method for entrapment of plasmids in liposomes. *J. Clin. Biochem. Nutr.* 1989; 7: 175-83.
3. Koshizaka, T. et al. Novel liposomes for efficient transfection of β-galactosidase gene into COS-1 cells. *J. Clin. Biochem. Nutr.* 1989; 7: 185-92.

ETHICAL REFLECTIONS ON HUMAN GENE THERAPY
Towards the Formulation of a Few
Questions and Some Answers

Paul Gregorios[*]

Two preliminary observations

As one looks at the ethical issues raised by new possibilities in human gene therapy, two fundamental observations emerge. First, gene therapy on human beings is still only beginning. So far, apart from a few bone-marrow transplants and some first steps in transfusion of blood (containing germ cells) from the umbilical cord of a newborn sibling, very little has been clinically established, so far as I know. The laboratory or experimental work which should precede clinical trials has barely begun, and still awaits ethically controversial, technically difficult and bureaucratically complicated decisions. Second, distinctions should be made, not only between somatic-cell correction in the patient and germ-line correction (ovum, sperm and embryo), but also among technical problems to be solved, ethical issues to be settled, and bureaucratic arguments to be overcome. A brief history of recent developments will make abundantly clear the significance of this third factor.

A brief history of some recent developments

In October 1984, Dr W. French Anderson, Chief of the Laboratory of Molecular Hematology at the US National Institutes of Health, a pioneer in this field as well as a prime actor in the dispute that ensued, reported in *Science* that gene-therapy experiments had been successfully completed in fruit-flies *(Drosophila melanogaster)* and in mice[1]. This was probably in response to a report in *New Scientist* of September 27, 1984 that human gene therapy was still a "distant prospect"[2]. Following Anderson's announcement, Stephen Budiansky suggested on November 29, 1984 that clinical trials of human gene therapy were just about to begin in the USA[3]. He did not know at the time that it would take more than four years to begin.

The US Congressional initiative in the matter was taken by Representative (later Senator) Al Gore (Democrat, Tennessee), through a bill asking the Congressional Office of Technology Assessment (OTA) to conduct an analysis of the ethical and scientific issues involved in human gene therapy. The OTA report appeared in December 1984; it made clear the distinction between somatic-body-cell correction in the individual patient and germ-line-therapy (ovum, sperm, embryo), which affects progeny as well. Gore, who had become senator by then, saw no unique ethical obstacles to the former, provided it met safety, efficiency and other standards as in any other therapy. The OTA report was hesitant, however, about the ethics

---

[*] Metropolitan of Delhi, The Delhi Orthodox Centre, New Delhi.

of germ-line genetic therapy: about the "need, technical feasibility or ethical acceptability of gene therapy that leads to inherited changes"[4].

The United States Congress has its own Biomedical Ethics Board, but that board has enough controversial issues to deal with, such as fetal abortion, *in vitro* fertilization and fetal research. It could meet only in December 1988 for the first time, and gene therapy was not on its agenda.

Congress has also a Recombinant-DNA Advisory Committee (RAC), consisting of 17 scientists, including Anderson and other gene-therapy researchers, and eight members of the general public. It has a special sub-committee on human gene therapy, chaired by Professor LeRoy Walters, Director of the Center for Bioethics at the Kennedy Institute of Ethics, Georgetown University, and consisting of three laboratory scientists, three clinical practitioners, three lawyers, two public policy specialists and one lay person. The subcommittee has first to approve gene-therapy experiments or clinical tests.

On August 15, 1985 Tim Beardsley reported in *Nature* that disputes between the National Institutes of Health and the U.S. Food and Drug Administration were likely to delay research into gene therapy[5]. Even in May 1986, neither the technical nor the ethical problems had been solved. Jean Marx reported in the Research Column of *Science* about the delay under the title: "Gene Therapy — So Near and Yet So far Away" and the sub-title: "Genes transferred into cultural cells often work very well, but getting good expression of foreign genes is another matter". Her bottom line was: "We are basically making the observation that we have a long way to go"[6].

It was only on 19 January 1989 that the first Federal approval for gene transfer was given. Even then, it was not a genuine case of gene therapy; the transplanted gene in itself gives no therapeutic benefit; it serves as a marker to track the progress of a promising but experimental cancer treatment. The experiment was to insert marker genes into tumour-infiltrating lymphocytes (TIL), or TIL cells, by means of a retrovirus incapable of replicating itself[7]. However, knowledgeable people thought that the last hurdle to somatic-cell gene therapy in humans had been overcome. Leslie Roberts reported in *Science* on 3 March 1989: "On January 19, the United Sates crossed the threshold into the much debated but still uncharted world of human gene therapy"[8] Though this was 17 years after the first gene had been spliced, Roberts had spoken too soon. On 30 January 1989, Jeremy Rifkin, President of the Foundation on Economic Trends (Washington, DC) accused the Recombinant-DNA Advisory Committee of ignoring the social and ethical ramifications of gene therapy. He called for a moratorium on the cancer experiment and on all human gene-therapy research until the NIH set up a special committee to evaluate the ethical and social implications of this work. His lawyer filed a suit in court to stop the experiment.

Meanwhile a new corporation, Genetic Therapy, Inc., had been set up to exploit clinically the newly emerging technology. Two of the NIH research scientists, French Anderson and Michael Blaese, had concluded a

Cooperative Research and Development Agreement (CRADA) with this corporation. They had agreed to keep the research data confidential and the company had agreed to pay them royalties if they kept their side of the agreement. Clearly there was more than one vested interest in the project.

The Federal Court eventually refused Rifkin's suit. The way was now cleared for the cancer experiment, which was carried out on 22 May 1989, but not by the two signatories to the CRADA agreement with Genetic Therapy, Inc. Their collaborator, Steve Rosenberg, who had refused to sign the CRADA agreement with the new company, finally carried out the implantation of the marker gene to trace the progress of the TIL cells introduced into the cancers of six terminally ill patients. Scientific journals heralded this milestone in medical history with catchy but cautious captions such as: *Treatment with genes creates medical history*[9]; *Gene therapy prelude under way*[10]; *Progress toward human gene therapy*[11].

The results of the experiment were regarded as moderately successful. Barbara Culliton's report in *Science*[12] (September 22, 1989) was captioned "Gene Transfer Test: So Far, So Good". Genetic-therapy research has now received a new push in the West as a whole. Several new clinical trials can be expected in the fields of gene therapy for various immune deficiency syndromes, chronic hepatitis C, hereditary cancer (through TIL techniques), cystic fibrosis and even AIDS (through molecular sabotage or the "mutant mimicry" of "throwing a monkey wrench into the genetic machinery of AIDS", as Rick Weiss reported in *Science News*[13]).

What about other countries? I am in no position to give anything like a comprehensive picture, and it is hoped that a lot of new information will emerge in this conference about what is going on in other countries.

The Government of the United Kingdom set up only in December 1989 a committee to investigate the ethics of gene therapy[14]. Sir David Weatherall, Nuffield Professor of Clinical Medicine at the University of Oxford, serves on this committee. UK Health Minister, Virginia Bottomley, wanted it to draft ethical guidance for doctors who might consider applying gene therapy.

The Government of Sweden has set up a Recombinant-DNA Advisory Committee, which has already organized in Stockholm (March 11 to 14, 1990) a conference on Trends in Biotechnology.

The European Commission has a $17 million project for Research in Human Genetics in 1990-91.

Many other projects and activities are expected to be launched in this field all over the world.

The ethics of genetic engineering — some earlier discussions

The science of genetics is new, but its rudiments had already been grasped in the last century and even then the possibilities implied in that limited knowledge raised an ethical discussion on eugenics, or selective breeding of humans (not genetic engineering as such, because it was not then possible).

145

The discussion started off with great furor as early as 1865, when Francis Galton, a younger cousin of Charles Darwin, wondered in writing, apropos of the then new discussion of possibilities of selective breeding of plants and animals for the benefit of farmers:

"Could not the race of men be similarly improved? Could not the undesirables be got rid of and the desirables multiplied?"[15]

Galton coined the word "eugenics" in 1883, to denote an almost religious doctrine, that the human stock could be improved by giving "the more suitable races or strains of blood a better chance of prevailing speedily over the less suitable". Daniel J. Kevles, in his "Genetic Progress and Religious Authority: Historical Reflections"[16], tells us that Galton proposed eugenics as a substitute for orthodox Christian doctrine. Clerical writers vehemently opposed Galton, but at the turn of the century, when the work of Gregor Mendel was discovered and publicly acknowledged, the "religion" of eugenics received a new boost.

In the period from 1900 to 1930 the "religion" of eugenics spread widely in England and in the USA, among liberals dissatisfied with traditional Christianity and charmed by the dramatic achievements of modern science. Eugenics itself became a sort of organized religion, and many organizations were formed for its promotion. A great deal of material was published on the subject. The American Eugenics Society was formed only in 1923, but already before its formation "Fitter Families Contests", begun at the Kansas Free Fair in 1920, were being repeated in seven or more States in America. In 1923 the quasi-religion of eugenics also published its Ten Commandments, *The New Decalogue of Science,* written by the most ardent evengelist of eugenics, Albert E. Wiggam, a journalist and popular lecturer. The book soon became a best seller. Three years later, the American Eugenics Society published its official catechism, *A Eugenics Catechism*[17]; it taught that eugenics could improve the human race and increase the number of geniuses by "selective love-making".

The gospel of eugenics received support from what were coming to be called "modernist" clergy, including Dean Inge of St. Paul's Cathedral in London. Liberal rabbis and Methodist pastors began using pulpits to propagate it, in America and Britain. In 1929 the Kansas Free Fair had a placard which asked in bold lettering: "How long are we Americans to be so careful for the pedigree of our pigs and chickens and cattle, and then leave the ancestry of our children to chance, or to 'blind' sentiment?". Our famous intelligence quotient (I.Q.) tests were first devised by a devotee of this religion, Henry H. Goddard, a psychologist who intended to use them as a means of detecting the "feeble-minded" and eliminating them to improve the human stock.

Why then did the movement suddenly die out in the 1930s? For two reasons: First, people began to see how it was becoming more popular among the WASPs (White Anglo-Saxon Protestants) of America, who wanted to keep non-WASPs out of power, and who feared that too many of these new immigrants were coming from 'inferior' races and blood-stocks, and Jews and Catholics at that! As people began to see hidden racist motivation behind the gospel of eugenics, it lost its popularity. The second reason for its failure was the use that Hitler began making of it in his doctrine and practice of National Socialism. We should be grateful to racism and Fascism for making the gospel of eugenics unpopular! Otherwise some of the less endowed among us in terms of I.Q., the more "feeble-minded" among us, would have lost the chance of having progeny; if eugenics had become universal, that could have happened to my parents, and I might not have existed! But eugenics is not the theme of this conference, and it does not even need the new genetic technology. I am sure, however, that eugenicists can find many uses for genetic engineering.

Closer to our own time and theme, in June 1983, a group of 59 prominent clerical leaders in America published and released to the press a "resolution" urging that "efforts to engineer specific genetic traits into the germ line of the human species should not be attempted". The signatories were all noted religious leaders, Roman Catholic, main-line Protestant, sectarian Protestant, Jewish and all, including Professor J. Robert Nelson, then of Boston University; Methodist Bishop James Armstrong, President of the National Council of Churches of Christ in the USA; Bishop Crutchfield, President of the Council of Bishops of the United Methodist Church; Dr Avery Post, President of the United Church of Christ; and Jerry Falwell of the Moral Majority.

This clerical outburst had been prepared for by similar explosions in the scientific community itself. In 1973 the International Congress of Genetics had raised quite a furor about the dangers inherent in the new technology of genetic engineering. The scientists were at that time prone to overestimate the possibilities for altering human personal characteristics and traits through genetic technology, at both somatic and germ-line levels. In 1974 a group of 11 of the world's leading geneticists called for a moratorium on recombinant DNA techniques for creating new bacterial strains. Were they justified in raising all that hullabaloo at the time? Well, who knows whether our common enemy, the AIDS virus, came into being as a result of some of these research projects to create a new virus to be used in war, and against which the human organism would have no immunity?

In 1975, 140 gene scientists assembled at the Asilomar State Park in California and issued guidelines for controlling or regulating research in recombinant DNA technology. In that same year a Harvard biologist had claimed to have synthesized a rabbit gene. In 1976 Har Gobind Khorana, a scientist of Indian origin, claimed to have made, entirely out of shelf chemicals, a gene that would work when implanted in a bacterial cell.

Some genetic scientists were already talking about using genetic technology at the germ-line level, not only to improve the human stock, but also to create new types of inferior human beings, organic robots who would do our dirty work for us, drones who would neither protest nor strike, workers who would not even ask for higher wages. Why not do so, if humanity is to take charge of its own destiny and reconstruct its own organic endowment, to suit its tastes and to meet felt needs? This is the question that scares, and probably made the eminent clergymen somewhat squeamish about genetic alteration of the human germ-line.

What is gene therapy?

In 1984, Dr French Anderson defined gene therapy as "the insertion into an organism of a normal gene which then corrects a genetic defect"[1]. Today we need a slightly broader definition.

Gene therapy should include diagnosis of genetic disease, genetic pharmacology (and pharmacopoeia), the multifactorial etiology and therapy of such major diseases as cardiac ailments, cancer, diabetes, arthritis, Alzheimer's disease and mental illness, in which proclivity to the disease, but not the disease itself, may be genetically inherited. Some 4,000 diseases are reported to be caused by genetic defects. However, in some, such as those listed above, genetic endowment provides only some proclivity to the disease ("it runs in the family"), while its onset may be determined by lifestyle and environmental factors. Clear diagnosis of the genetic endowment may help in adopting a prophylactic lifestyle and environment as part of the therapy.

Diseases of non-genetic origin can be genetically treated, it now seems. Intracellular immunization by gene therapy is no longer a purely theoretical question. Interferon in cancer therapy is a clear case. Experiments are now in process on a technique of biomolecular sabotage, to which earlier reference has been made. The idea is to alter genetically a patient's white blood cells, stimulating them to manufacture a constant supply of mutant viral components. An invading HIV virus is supposed to contact these mutants and lose its ability to carry on the replicative cycle. This research is now reported as in progress in at least three places: Duke University Medical Center (Durham North Carolina), the St. Louis University School of Medicine, and the National Institutes of Health, at Bethesda, Maryland[13]. If this proves successful, the present emotional furor against gene therapy could subside substantially, as AIDS is brought under control for the first time.

Human gene therapy thus cannot be narrowly defined as replacing a "defective" gene by a "normal" gene. It must include any genetically engineered alteration of the human constitution (somatic or germ-line), as well as genetically produced "medicines" and mechanisms, and the techniques used for gene transplant — retroviruses, endothelial cells, lymphocytes and neo-organs or genetically created and implantable organoids[18].

There are ethical problems involved not only in replacing one gene by another but also in the techniques and methods used to effect such replacement.

There is another question to be raised at this stage. Since the environment plays a partner role in the onset of disease along with genetic endowment, should we include in genetic therapy also genetically engineered modification of the human environment? Western medicine is now at the initial stage of the process of moving away from body-centred therapy to a more holistic therapy, which includes psychosomatic, social and environmental factors. We shall leave this question for the time being.

The nature of ethical reflection

The word "ethics" is of ambiguous meaning. Let us think of it as reflection on the norms of human behaviour. The problem begins when we seek criteria for validation of ethical norms. These norms are not, objectively given, but derived from previous human experience, which includes religious experience and participation in a religious community.

Anglo-American philosophers used to argue that no "ought" can be derived from an "is". In other words, objective facts do not provide norms, according to them. This argument has been called in question. The statement that no "ought" can be derived from an "is", though couched in propositional form, is itself an "ought" statement. What then is its philosophical validity?

In fact, some philosophers argue the exact reverse, that every "is" is derived from an "ought". When we see a Japanese kimono, for example, we cannot say that this is a kimono unless we have some previous knowledge of what a kimono ought to look like. The "ought" is already in our mind, which is why we can state the "is".

There is a third philosophical position, which argues that every "ought" is derived from an "is". The basis of this position is that all ethics is ultimately grounded on some affirmations, religious or secular, about the nature of reality. If one believes that humanity is created in the image of God, then certain ethical norms naturally flow from that proposition. If, on the other hand, one's belief is that humanity is naturally evil and cannot but do evil, the ethical consequences are different. Let me state quite clearly from the outset that my position is the former, not the latter. I believe that humanity is created in the image and likeness of God. This does not mean that God is corporal or mortal. It does mean that the most essential things about human beings is to be free, good, wise, loving and creative, like God.

If, however, one believes that humanity is a chance or accidental product of blind evolution, the ethical values may be derived from such concepts as "survival of the fittest", which includes elimination of the unfit. I submit for your consideration that it is difficult to derive some of our more lofty ethical concepts from a purely secular framework: human rights, justice for all, democracy, or the freedom and dignity of all human beings and the unity of all humanity, to cite four of the major values which many of us regard as vital to the civilization of the future.

The ethics of human gene therapy

Some of the popular animosity against gene therapy on humans comes from unexamined assumptions. Since Hiroshima and Nagasaki many have acquired a bias against modern technology as such, that it is basically anti-human and destructive. The power of the military factor in our societies today reinforces this bias. Our persisting inability to eliminate weapons of mass destruction confirms the suspicion that modern science/technology does not easily come under beneficent human control. Therefore, people are often apprehensive about scientists and professionals acquiring more and more technological and financial power.

There need not be any doubt that advances in gene therapy will give the specialist added power to intervene in other people's bodies. However, this has been the case with every advance in medical or surgical expertise, and there cannot be anything intrinsically evil about the process.

Some people argue that altering the "God-given" genetic endowment or a person or an embryo is interfering with God's work or nature. They can rhetorically refer to such intervention as "playing God". If gene therapy is thus "playing God", then building a dam or shaving a man's face should also be interfering with the way God has arranged things.

Some of us from the Eastern Orthodox Christian tradition would argue that most of humanity's task in this world is to act as the presence of God, through freedom and creativity, love and goodness, wisdom and power, prayer and activity, to transform the whole world into something reflecting the glorious goodness of God. From that perspective one cannot see anything intrinsically forbidden or evil in gene therapy, whether somatic or germ-line. Infinite possibilities of power are open to humanity. The ethical problem is not in the acquisition of this power, but in its wise use.

Actually, all power, whether technological or other, in a free being, is subject to this ethical requirement, namely that it be used wisely, in the interest of what is good — good not just for oneself, but good for all. Ethics is in fact reflection about the question: what is good for all and how do we discern that good in a particular situation?

Let us put our heads together on a few hypothetical cases. Lacking technical competence to judge whether a hypothetical possibility is technically feasible or not, I feel rather free to hypothesize.

Supposing criminal tendency in a hard convict is due to some chromosomal defect, and supposing the technology exists to correct the defect, would it be ethically justifiable to apply such therapy? Obviously you would agree with me that the prior consent of the convict, who now turns out to be a patient rather than a criminal, would be a desirable component of the ethical package. But would a public court be justified in sentencing a convicted criminal to genetic correction and rehabilitation rather than a death sentence or life imprisonment or similar heavy penal measure? Would such a sentence be an encroachment on human rights? To me it seems it would be no more a violation of human rights than execution or imprisonment. You may disagree.

At the moment we do not have enough practical experience in human gene therapy to know what its side-effects might be. Suppose some gene therapy that corrects some somatic or germ-line defect brings on a syndrome which we do not know how to handle clinically; we would then have to do a new cost-benefit analysis before deciding whether such therapy is desirable. Some such vague apprehension is in the unconscious minds of many who have seemingly irrational fears about gene therapy. I myself suspect that questions of this kind cannot be handled *a priori*. A certain amount of clinical experience alone can adequately expose the ethical and technical problems in this line.

An important ethical question will always be the cost/justice factor. It seems obvious now that gene therapy is unlikely to be inexpensive. This will put such therapy beyond the reach of large numbers of people. Such is already the case with many aspects of diagnostic and therapeutic medical technology. How far can we ethically justify using public funds for research intended to reap benefits not available to all who need them?

Using the broader definition of gene therapy as including prenatal as well as postnatal genetic diagnosis and screening, should we make genetic screening compulsory for all pregnancies, whether at public cost or at individual expense? A question like this needs a great deal of public discussion, especially by the taxpayers.

Statistics indicate that 5% of all infants born in the U.S.A. carry genes that make inevitable some form of sickness or suffering — pain, crippling, mental retardation, severe disability, early death, the onset of a disease such as Huntington's chorea in late adulthood, and so on. What are the options after prenatal detection? Gene therapy is not likely soon to advance to the stage where prenatally detected genetic diseases can be treated *in utero*. The other possibility is abortion of the embryo, and this raises a host of ethical questions.

First, there is the issue of the legitimacy of any kind of abortion in any circumstances. Here cultures with a long history of martial killing may find it easier to make decisions in favour of abortion whenever needed. Also cultures where the individual is supreme can argue about a woman's rights over her own body. Other cultures may want to hold that abortion is killing and cannot be justified at all. This is the Eastern Orthodox Christian position, adopted by St. Basil the Great in the 4th century. In practice, however, even more conservative cultures sanction abortion of the fetus if the mother's survival is imperilled. However, when a genetic defect is prenatally diagnosed, the option is first whether to have the baby anyway or to abort the fetus. Who should make this decision: parents on their own, the physician alone, or the two together, with some consultation with a pastor in the case of religiously oriented parents?

Second, there is the question of what constitutes a "defective" gene. Defects are measured against some given standard. To be blind or lame constitutes a defect or handicap for a human being, but it does not certainly make such a person less than human. Why should a genetic defect make a person less than human and deny that person the right to live? Once the principle is adopted that a genetically defective person is less than human and therefore does not deserve to live, the logical conclusion would have to be drawn that all persons now existing should, without exception, be genetically screened, and defective persons put to death.

As for germ-line therapy, people have more complex apprehensions, not about the first test-cases, but about how eventually the technology will be used. Would governments oblige people carrying defective genes to undergo gene therapy in the interest of so-called positive eugenics? Would laws emerge requiring gene therapy before such people could have children?

Genetic screening may bring out information about genetic disease or predisposition to disease, which an employer or an insurance company can use against the individual.

The gap, likely to persist, between genetic diagnosis and gene therapy creates another ethical issue. Why should we make so much progress in genetic screening before we have made more progress in research on how to cure genetic defects?

Even more important is the question: who should fund and control research into gene therapy? Should we leave it to private enterprise? If it is controlled by a bureau, scientists fear that such control would cripple progress. Others feel that since genetic technology touches the deepest sources of human identity, there should always be an extended public discussion before important research is undertaken. The consequent delay costs less than the risk involved in developing new genetic therapies and techniques, some would say.

Would social opprobrium be attached to gene therapy, as in the case of mental illness? To me this seems a minor ethical consideration, which should not stand in the way of research into gene therapy.

The real fear, deep in the minds of people, relates to the encroachments of technology on the human person. We have such a high regard for our human genetic constitution that we are not sure about allowing technology to interfere with it. This uncertainty is not totally irrational: it is based on past experience in which technology has damaged our human environment, about which we previously knew so little.

Uncertainty demands caution, but does not mandate inactivity. Some things humanity learns only the hard way, by trial and error, and by such colossal errors as the development of weapons of mass destruction. In some cases its seems almost impossible to undo the damage once the error has been made. There was no public discussion about the rights and wrongs of atomic and hydrogen bombs before they were made. We should be grateful that discussion on gene therapy is now under way, before its clinical use goes very far.

References

1. Anderson, W.F. Prospects for human gene therapy. *Science* 1984; 226:401-9.
2. Gene therapy is still a distant prospect. *New Scientist* 1984 Sep 27; 103:15.
3. Budiansky, S. Gene therapy: U.S. clinical trials imminent. *Nature* 1984; 312:393.
4. Baskin, Y. Doctoring the genes. *Science* 1984; 5:52-60.

5. Beardsley, T. Gene therapy: NIH/FDA dispute likely to delay research. *Nature* 1985; 316:567.
6. Marx, J. Gene therapy — so near and yet so far away. *Science* 1986; 232:824-5.
7. Roberts, L. Human gene transfer test approved. *Science* 1989; 243:473.
8. Roberts, L. Ethical questions haunt new genetic technologies. *Science* 1989; 243:1134-6.
9. Joyce, C. Treatment with genes creates medical history. *New Scientist* 1989 May 27; 122:29.
10. Marwiur, C. Gene therapy prelude under way. *JAMA* 1989; 262:16.
11. Friedmann, T. Progress toward human gene therapy. *Science* 1989; 244:1275-81.
12. Culliton, B. Gene transfer test: So far, so good. *Science* 1989; 245:1325.
13. Weiss, R. Mutant mimicry. *Science News* 1990; 137:43.
14. Britain examines ethics of gene therapy. *New Scientist* 1989 Dec 9; 124:22.
15. Pearson, Karl: The Life, Letters and Labours of Francis Galton, Cambridge, Cambridge University Press, 1914/1930. Cited by Daniel J. Kevles: "Genetic Progress and Religious Authority: Historical Reflections" in Byrne, K. ed., *Responsible Science: The Impact of Technology on Society*, Harper and Row, San Francisco, 1986. pp. 31-48.
16. Kevles: op. cit. see note above.
17. American Eugenics Society, *A Eugenics Catechism*, 1926.
18. Schwartzenberg, P.L. *et al.* Germ-line transmission of c-abl mutation produced by targeted gene disruption in E.S. cells. *Science* 1989; 246:799-802.

GENETIC THERAPY — POLICY-MAKING ASPECTS
A Parliamentarian's Perspective

A. Inayatullah[*]

1. Overview

1.1 Only four decades ago, human genetics was considered a thinly cultivated field. Human genetics could be summed up as consisting only of four blood groups and red-green colour-blindness as simply inherited characters of anthropological relevance. Since then, however, considerable progress has been made in our understanding of human biology, which includes human genetics. This has been due largely to the elucidation of the structure of deoxyribonucleic acid (DNA), and the discovery of its remarkable capacity for encoding and passing on genetic characteristics from one generation to the other. These are post-1953 developments. The rapidity with which the field has grown is startling. It is now possible to isolate specific DNA sequences from one species and attach this genetic material (also called recombinant DNA) to a different species. Essentially, by this technique, also called genetic engineering[1], it is possible to add genetically determined characteristics to cells that would not otherwise have possessed them. There are also other means of altering the genetic pool — for instance, *in vitro* fertilization and cloning[2]. Neither *in vitro* fertilization nor cloning necessarily involves genetic manipulation, although both techniques might be used in conjunction with such manipulation in particular circumstances.

1.2 The rise of genetic engineering has brought immense new possibilities for human welfare. Recombinant DNA technology (gene splicing) has made it possible to produce a number of drugs and biologicals, for example: (a) human growth hormone for the treatment of dwarfism; (b) human insulin for the treatment of diabetes; (c) urokinase, an enzyme which dissolves blood clots and may be useful in treating thrombosis; (d) anti-hemophilic factor, which is essential for blood clotting and is used to treat hemophilia; (e) interferon, which could curtail the spread of certain cancers (this natural product is now being produced through gene splicing and is being tested in clinical trials); (f) vaccines for the control of polio virus[3] and hepatitis B virus[4], and antibodies against surface antigens (recombinant technology). Another development is cell fusion, by means of which monoclonal antibodies can be harnessed in the fight against cancer and "oncogenes", which direct the wide proliferation of cells that creates a tumour, can be identified and controlled.

1.3 An interesting spin-off of recombinant DNA technology employs the specificity of restriction enzymes to help diagnose the existence, or carrier nature, of a wide range of genetic disorders that until now have not

---

[*] Chairman, Family Planning Association of Pakistan, Lahore.

been readily diagnosable[5]. This tehnique, known as genetic screening and diagnosis, holds particular promise for prenatal tests and for diagnosis of late-onset disorders such as Huntington's disease - a degenerative and fatal disease of the nervous system, determined by a single dominant gene, and appearing on the average at about age 35.

1.4 In the immediate future the most important applications of gene splicing to human health will probably be in the creation of drugs and biologicals, vaccine production and genetic screening. Also, the use of technology for curing human genetic disorder is making rapid progress. The development of this phase is to be watched with interest, for, in so-called "gene surgery", either the defective gene is excised or its functions are suppressed, so that it no longer would be in a position to send out a message that would result in a defective product.

1.5 With rapid increase in the scope and application of genetic screening and gene surgery it is becoming increasingly easier not only to detect a genetic disorder at an appropriate stage of the human reproductive cycle, but also to offer a precise cure through introduction of normal genes in place of those transmitted through error in the chromosomes. These advances have rendered the task of genetic counselling even more exacting.

1.6 In essence, then, man today is in a position to diagnose genetic disorders, treat some of them by means of gene surgery, and provide genetic counselling even in the cases of those genetic disorders that can be detected at different stages of pregnancy.

1.7 While commending these developments in genetics a word of caution must also be introduced: they must be monitored and their associated risks kept under surveillance. Their application should be adapted to the social and cultural environments in which it occurs. Moreover, in view of the known costs, priority should be given to introducing in developing countries what is today in developed countries simple and tried technology.

1.8 The emergence of the science of genetic engineering combined with *in vitro* fertilization and cell fusion has raised several ethical and social issues. In particular, the General Secretaries of the National Council of Churches, the Synagogue Council of America and the United States Catholic Conference have drawn attention to the associated moral, ethical and religious questions, in a letter to the President's Commission (USA) for the Study of Ethical Problems in Medicine and Biomedical and Behavioral Research, and contained in the Commission's report, *Splicing Life*, on the social and ethical issues of genetic engineering with human beings[6]. Interestingly enough, such concerns had led in 1973 and 1974 to the publication, in both *Science* and *Nature*, of letters signed by several leading microbiologists. The first letter called attention to the issues and asked the National Academy of Sciences (USA) to establish a committee to investigate the social and ethical aspects of genetic engineering. The second urged the scientists to hold off on certain DNA recombinant experiments until the risks could be assessed[7-9]. It was thanks to the President's Commission[2], and more so to the Asilomar Conference[9], that the self-imposed moratorium was lifted for most recombinant DNA research, subject to specified physical and biological containment measures that would be graduated according to the risk of the experiment. The moral and ethical issues pertaining to genetic screening, gene surgery and genetic counselling are the subjects of several reports[6,10]. The mechanisms of treating genetic disorders — in particular, those that use "gene surgery" or even gene therapy — raise the fundamental question of man's interference with nature. Its implications are much wider than simple medical treatment of a disease and in turn have policy implications.

1.9 The benefits of gene surgery and genetic screening are obvious. The costs are high, however. Perhaps the current technology, as well as new technology, will remain beyond the reach of a large segment of the world's population, unless supported by public resources. This negates the principle of equality of opportunity. Should we not ask: who will benefit from the new technology? Will the costs and benefits be distributed equitably[2]?

1.10 This aspect raises two important questions. First, to be born with a genetic defect is not what a person has earned, but is the result of a "social lottery"[11]. At some time in their reproductive life, about half the world's women conceive a child with a chromosomal abnormality. The supporters of equality of opportunity have urged that social institutions be designed so as to minimize or compensate for the influence that the "social lottery" exerts upon a person's opportunities. Does this run against the fundamental principle that distributive justice deals with inequalities in social goods and plays no role in regulating natural inequalities? Second, what about the wide differences between the developed and the developing countries in economic resources and levels of achievement of technology? The disadvantages at which the developing countries are placed preclude the possibility of effective use of these technologies, or even programmes of genetic counselling and genetic screening. Because of poor standards of health and education in these countries, and scarce knowledge of the new methods of genetic screening, inequalities between the developed and developing countries are likely to be perpetuated. This should be a cause for concern for those who believe in equality of opportunity, and who, above all, view the balanced growth of the human species in all parts of the globe as an international obligation.

1.11 This calls for practical suggestions. Perhaps it would require the setting up of a fund for expanding training programmes and for the rapid transfer of technology, if genetic screening is to become a part of maternal and child health services in developing countries. At present, for instance in such countries as Pakistan, nothing of the sort exists. Consanguinity is common. Family after family amongst the rural poor is producing and rearing genetically defective children, thereby increasing the flow of defective genes from one generation to another. One illuminative instance of a pedigree with stenodactyly in rural Sind (Pakistan) has recently been worked out. Perhaps we need to pay more attention to such populations? Perhaps we need to arrange genetic counselling and screening services for them.

In the developing countries we do not know about even the endemic prevalence of a large number of genetic disorders, for which elaborate services have been developed in a number of such countries. As the world is shrinking, it becomes more important for the developed countries to know about the distribution profiles of genetic disorders in world populations.

1.12 It is important to determine who will decide upon the extent to which genetic surgery or therapy could be used, the purposes for which it would be used, the circumstances and the consequences. This calls for regulatory controls, which perhaps could best be instituted by an international commission on the subject, with the reports of the President's Commission for the Study of Ethical Problems in Medicine and Biomedical and Behavioural Research[2] serving as guidelines.

1.13 The commercial value of new technologies of gene surgery and gene therapy, and also of genetic screening by recombinant DNA or hybridization, is creating a different type of university researcher. This is taking toll of the advancement of basic knowledge in the universities. How will this trend affect academic research? Is there a need for a new and liberal policy of funding basic research in the universities? These questions need to be addressed squarely in the perspective of their putative academic role.

2. Policy implications

From the foregoing discussion it is recognized that genetic screening and genetic therapy have a potential for human well-being.

2.1 A growing number of genetic diseases can now be accurately diagnosed and genetic information of potential value provided to individuals and families. Every couple is at risk of having an abnormal offspring. Some recent estimates indicate that 50% of the world's women conceive a child with a chromosomal abnormality. Also, most infants with congenital malformations and chromosomal disorders are born to healthy women with no previously identifiable risk factors. The occurrence of these sporadic disorders cannot be prevented, nor is intervention possible before the women become pregnant. Thus, it is only by means of well-designed and carefully implemented genetic screening and counselling programmes that individuals can have the opportunity to make informed autonomous decisions about their own health and about reproduction. Indeed, at present there is little disagreement about the major contributions which genetic screening and genetic counselling programmes can make to public health and personal well-being by reducing the incidence of genetic diseases.

2.2 Given the widespread occurrence of genetic diseases and the fundamental importance of prenatal diagnosis, cheap, accurate and safe mass-screening services should be made available to women who are prepared to accept genetic screening. Also, genetic screening and genetic counselling programmes must be dovetailed. Genetic counselling is an expanding and evolving field, the value of which will increase as more genetic tests become available. Screening and counselling, performed according to sound ethical and public-policy principles, should certainly enable couples, particularly those unlikely to change partners, and who understand the risks, to use whatever means are available to them to avoid the birth of severely affected children.

2.3 Both genetic screening and genetic counselling programmes, because of their ever-increasing role in medical care, are associated with several social, ethical and legal issues. Generally, the blossoming of medical genetics in the 1960s drew increased public and professional attention to the benefits of genetic screening and genetic counselling, but also to their ethical implications. Some of these concerns were unique to genetic screening and genetic counselling, while others were familiar issues, such as those raised generally by biomedical research involving human subjects and abortion[12,13].

2.4 The main ethical principles include autonomy, beneficence (including the prevention of harm), distributive justice (including equality of opportunity and fairness), and confidentiality.

2.5 The main public-policy areas of concern are:

Efficiency (or economy)
Public participation (through democratic political institutions)
Inequality (equality of opportunity)
The role of the State (in protecting well-being and respecting individual liberty)
The education and role of the counsellor (directive versus non-directive advice)
Public and professional education
Evaluation
Equity (use of some techniques only for advanced maternal age)
The authenticity of tests (avoidance of risk, and who decides the use of a test)
Levels of operation (special service — clinical, fetal medicine; community service through primary health care system)
Selection of a method
Selection of a method of choice (e.g. ultrasound) and the association of clinical genetics centres with expert fetal-medicine centres
Support for research laboratories (Third World? Biotechnology, Laboratory services)
Option of selective abortion
Improved education in genetic methods
Technical training of medical and non-medical staff
Accessibility of genetic counselling

2.6 Any discussion on social and ethical aspects of genetic counselling should consider confidentiality and autonomy, and disclosure of information to unrelated third parties such as employers and insurance companies. The information stored in data-banks, and disclosure of information to relatives, call for appropriate safeguards. For example, information to a

158

third party or even to relatives should be provided only with the consent of the person concerned, and the information stored in data-banks should be coded. Also, as regards autonomy, the use of genetic services should be left to the choice of the user. However, there is still controversy about directive and non-directive counselling: the balance of evidence favours directive counselling.

2.7 The subjects of "equal opportunity" and "distributive justice" have been touched upon under Section 1.10 above. On the whole, the developing countries have inadequate or, in many cases, no genetic counselling or screening services. In several, consanguinity is very common. Latterly, the frequency of consanguineous unions has been declining in Japan[14]. However, high levels of inbreeding have been reported among the predominantly Hindu people of South India[15] and in traditionally Muslim countries, ranging from Egypt[16,17] and Kuwait to Sudan[18], Iran and the southern Republics of the USSR. Recent studies[19] undertaken on the population of seven Pakistani Punjab cities show that consanguineous marriages are strongly favoured; the coefficient of inbreeding (F) in these populations ranged from 0.0236 to 0.0286. A significant relationship was found between inbreeding and prenatal and postnatal mortality.

2.8 There is a whole range of policy implications for initiating, implementing and extending the programmes dealing with genetic disorders. Firstly we should be looking for cheap, easy and reliable methods of genetic screening; obviously with the advancement of biomedical technology, more knowledge-intensive techniques will enter into diagnostic programmes. Thus, to be able to cope with genetic disorders in world populations, especially in areas where little progress has been made in this regard, it will be necessary to accelerate the dissemination of information about genetic screening and genetic counselling, the consequences of consanguineous marriages, and the need to eliminate deleterious genes from populations. This can best be achieved through the operation of systematic and organized public information programmes among other media inputs; this would require, for example, the preparation of video-cassettes and their wide distribution, and the communication of the mysteries of our genes to the layman in simple language. Secondly, the subject of human genetics, taught in medical schools, will have to be updated with modern knowledge of dealing with genetic disorders. Thirdly, much of the new technology used in screening is research-based. Research structures are weak in developing countries. Therefore new initiatives are needed to mount a massive programme of transfer of technology. This could happen through training of personnel and supply of equipment to developing-country research and technology institutions. Fourthly, a working group should be established to work out the minimum equipment, space, facilities, and trained personnel required for setting up an efficient and effective genetic service centre. Such centres should be offered as a package and established as part of maternal and child health services. Genetic counselling should go hand in hand with genetic screening in these centres. A special fund would be needed for this purpose.

Such agencies as the United Nations Fund for Population Activities and the World Health Organization should be invited to assist in this important public health area.

In conclusion, it may be said that the last four decades have seen major advances in genetics. It is essential that the human-values and ethics base of public policy keep pace with these scientific developments. Among others, the following policy implications have been identified:

(1) Genetic disorders remain largely neglected and constitute an important public health problem. In line with the principles of the Alma-Ata Declaration, priorities should be directed at the screening and counselling of genetic disease at the primary-health-care level in developing countries.

(2) The indigenous approach, in which the use of technology is both appropriate and socially relevant, should be encouraged.

(3) New possibilities for affected parents have been opened up owing to the introduction of modern biomedical technologies in the diagnosis of genetic disorders. Clinical genetics needs to be incorporated in the specialist-obstetrics infrastructure.

(4) A world-wide programme of genetic screening and genetic counselling needs to be launched, which, among other things, recognizes that, because reproductive risk cannot be eliminated, prenatal diagnosis and the option of selective abortion will be needed.

(5) The developing countries with a poor research infrastructure need rapid transfer of technology and an internal readjustment to ensure a balance between basic research and commercial-value research.

(6) For equitable delivery, the funding of genetic services, to be introduced in the community by setting up centres as part of the maternal and child health system for diagnosis, counselling and treatment of genetic disorders.

(7) An international commission should be set up by WHO to establish regulatory controls on the judicious use of genetic screening — specifically, gene therapy — and to devise an agreed international code of ethics to control abuses of genetic knowledge.

(8) The social and ethical issues raised by genetic screening, gene therapy and genetic counselling should be examined and their national, regional and international policy dimensions addressed.

(9) For the equitable delivery of services and universal accessibility, emphasis should be placed on the development and use of cheap and accurate methods which would facilitate mass screening and increase accessibility of prenatal diagnosis and genetic counselling.

(10) With a view to transcultural application, a data-bank of genetic disorders should be established. This cataloguing of human genetics should make possible the distribution of information on genetic disorders that are detectable through mass screening. This should be based on regional cooperative efforts.

160

(11) The language of communication should be strengthened. Public awareness should be increased by means of education and information, reaching out to:
  (a) children, through the incorporation of simple genetic information in school curricula.
  (b) the community, through media materials and orientation of local leaders.
  (c) health-care staff, such as doctors, midwives, lady health-visitors and nurses, by training them in basic genetic counselling.
(12) Genes have no passport, they are the common heritage of man. Therefore, a special fund needs to be created for launching the above policy initiatives in developing countries.

In conclusion, it is evident from the breadth of the issues raised that biomedical sciences have reset the operative framework in a pluralistic world. However, one thing is clear: in accepting the right of subjective individual preference, scientists, ethicists and policymakers have the joint responsibility of applying the utilitarian principle.

References

1. Hotchkiss, D. Portents of genetic engineering. *J. Heredity* 1965; 56: 197.
2. *Report on the President's Commission for the Study of Ethical Problems in Medicine and Biomedical and Behavioral Research: Gene Splicing,* Washington, 1988 pp. 9-18.
3. Racaniello, V.R. & Baltimore, D. Cloned poliovirus complementary DNA is infectious in mammalian cells. *Science* 1981; 214: 916-8.
4. Blumberg, B.S. *et al.* The relation of infection with the Hepatitis B agent to primary hepatic carcinoma. *Am. J. Pathology* 1975; 81: 669.
5. Emery, A.E. Recombinant DNA technology. *Lancet* 1981; 2: 1406.
6. *Report on the President's Commission for the Study of Ethical Problems in Medicine and Biomedical and Behavioral Research: Splicing Life*, U.S. Government Printing Office, Washington, 1982.
7. Rogers, M. *Biohazard*, Alfred A. Knopf, New York. 1973. (pp. 51-101).
8. Bennet, W. & Curin, J. Science that frightens scientists, *The ATLANTIC* 1977; 43: 49-50.
9. Dworkin, R.B. Science, society and the expert town meeting: some comments on Asilomar. 51 s. *Cal. L. Rev.* 14/1 (1978).
10. *Report on the President's Commission for the Study of Ethical Problems in Medicine and Biomedical and Behavioral Research: Screening and Counseling for Genetic Conditions*, U.S. Government Printing Office, Washington, 1983.
11. Rawls, J. A Theory of Justice. Cambridge, Mass., Harvard University Press. 1973. Quoted in (6, pp. 68-9).
12. *Report on the President's Commission for the Study of Ethical Problems in Medicine and Biomedical and Behavioral Research: Protecting Human Subjects*, U.S. Government Printing Office, Washington, 1981.
13. *Report on the President's Commission for the Study of Ethical Problems in Medicine and Biomedical and Behavioral Research: Implementing Human Research Regulations*, U.S. Government Printing Office, Washington, 1983.
14. Imazumi *et al. Jap. J. Hum. Genet.* 1975; 20: 91-107.
15. Rao, P.S.S. & Inbaraj, S.G. *Ann Hum. Genet.* 1979; 40: 401-13.
16. Hafez M. *et al. J. Med. Genet.* 1987; 20: 58-100.
17. Al Amadi, S.A. *et al. Clin. Genet.* 1986; 29: 384-8.
18. Ahmed, A.H. *Br. J. Psychiatry* 1979; 134: 635-6.
19. Shami, S.A. *et al. J. Med. Genet.* 1989; 26: 267-71.

HEALTH POLICY ASPECTS OF GENE THERAPY

Walter P. von Wartburg*

Gene therapy — progress or promise?

Impressive progress has been made during recent times in understanding the pathogenesis of human genetic diseases. The tools of molecular biology have permitted the isolation of many disease-related genes and the hoped-for completion of a human genetic linkage map will probably accelerate the genetic characterization of a great number of genetic diseases.

The molecular characterization of human genetic diseases will increase markedly the number of those diseases that we recognize to have major genetic components. It will also permit the development of interesting new approaches to diagnosis, detection, screening and even therapy of disorders by aiming directly at the mutant genes rather than at the gene products. The preparation of a physical map of the entire human genome and the information that will flow from this effort are likely to affect fundamentally the management of many important genetic diseases.

Current therapies for most human genetic diseases are inadequate. In response to the need for effective treatment, modern molecular genetics may provide tools for a new approach to disease treatment by means that will attack mutant genes directly. Recent results with several target organs and gene-transfer techniques have led to a concept of gene therapy for disorders of the bone marrow, liver, central nervous system, and some kinds of cancer, as well as deficiencies of circulating enzymes, hormones and coagulation factors. The best-developed models involve an alteration of mutant target genes by gene transfer with recombinant pathogenic viruses in order to express new genetic information and to correct disease phenotypes.

Of the many techniques in use to introduce new genes into cells, some are of particular interest for gene therapy. **Transfection** has been the most extensively studied and has been used to put new genes into mice and men. **Homologous recombination** offers the promise of curing certain mutations *in situ*, although it is still largely unexplored. **Injection** of new genes into the nuclei of single cell embryos has had some success in animals, but there are many scientific issues to be resolved before such an approach can be considered for gene therapy. **Retroviral vectors** offer the most promising prospect for introducing useful gene sequences into defective cells. Safety issues and problems of low levels of gene expression are still to be resolved, however. It appears likely that the bone marrow will be the most promising target of gene therapy for genetic diseases. The marrow is accessible and

---

* Professor, Department of Law, University of St. Gallen, St. Gallen, Switzerland.

contains proliferating stem cells which can transmit the new genes to progeny cells.

There are four potential areas for the application of genetic engineering designed to insert a gene into a human being:

— *Somatic cell therapy*. This would result in correcting a genetic defect in the body cells of a patient.

— *Germ-line gene therapy*. This would require the insertion of the gene into the reproductive cells of the patient so that the disorder would also be corrected in future generations.

— *Enhancement genetic engineering*. This would involve the insertion of a gene to try to enhance a known characteristic of a person, such as placing an additional growth-hormone gene into a normal child.

— *Eugenic genetic engineering*. This would represent the attempt to alter or improve complex human traits which are coded by a large number of genes involving, for example, personality, intelligence, and formation of body organs.

Somatic-cell gene therapy is regarded as technically the simplest and ethically the least controversial form of gene therapy.

Germ-line gene therapy will require major advances in present knowledge, and also raises ethical issues which clearly need to be debated further.

Enhancement genetic engineering presents a host of difficult ethical concerns. Unless this type of therapy can be clearly justified on the grounds of preventive medicine, prevailing opinion suggests that enhancement engineering should not be performed.

Eugenic genetic engineering is still impossible and will probably remain so for the foreseeable future despite all the attention it continuously receives in the public and political arena.

The promise of gene therapy thus provides new options for medicine. However, like all progress, it also implies new responsibilities for decision-makers. They relate to those limitations which reason, morality and ethical considerations impose upon the manipulation of the human genome for the benefit of both patients and society.

Without wanting to arrive at clear-cut conclusions, the following sections are intended to deal with the health-policy aspects and legal implications of gene therapy in a conceptual way. They look at gene therapy as a new form of medical intervention, which will have to be discussed in political, economic and legal terms. There is a need for the elaboration of a public policy with respect to gene therapy, in the context of general public-policy considerations. Health-policy objectives will play a major role in deciding on the future role of gene therapy. Moreover, the factors that influence human health, and the role of government in influencing those factors, will also have to be considered when formulating gene-therapy policy. The role of government, with its functions of health protection, health promotion and health care, needs to be intensively discussed.

Gene therapy in relation to politics, economics and the law

Politics is concerned with defining and achieving objectives within a given society. The process of defining the objectives may be carried out collectively (as in a democracy) or individually (as in a dictatorship). Achievement of the defined objectives is possible on a centralized or a decentralized basis. The form of organization adopted and the values underlying the policy pursued may vary according to the political philosophy of the society and State concerned. The same applies to health policy.

Economics contributes to the processes involved in defining and achieving objectives by providing for rational alternative lines of action. Economics is concerned with questions as to which resources should be employed and in which combination (allocation of resources), by which methods goods and services should be produced (production), and how the goods and services made available should be distributed (distribution).

Economics, like the law, is hierarchically subordinate to politics, because its normative guidelines are of political origin. What constitutes, for example, a "just and fair" distribution of incomes is a political, and not an economic, decision. Economics, however, can indicate the tools with which the defined objectives may best be achieved. Economics in this context is thus an auxiliary science which not only aids politics but also exerts an appreciable influence on the political processes involved in defining and achieving objectives.

The **law** is responsible for ensuring that these objective-defining and objective-achieving processes are duly reflected in the realms of legislation, administration and justice. In the realm of legislation, the law can be regarded as embodying political guidelines. In the realm of administration, however, the law should already be largely independent of politics. Finally, in the realm of justice, the law stands beyond politics and, in the last resort, must adhere to the basic tenets of the constitution.

Gene therapy as an emerging new form of medical intervention has to be seen in the context of politics, economics and the law. In a political context societies and governments will have to decide on the strategies they wish to pursue with respect to making gene therapy available and controlling its application. Economic considerations will come into play when deciding about the pricing, the financing, the cost reimbursement and the general or specific availability of gene therapy. Legal rules and regulations will be necessary to carry out the political mandates and the economic decisions related to gene therapy in an open and transparent fashion (Fig. 1).

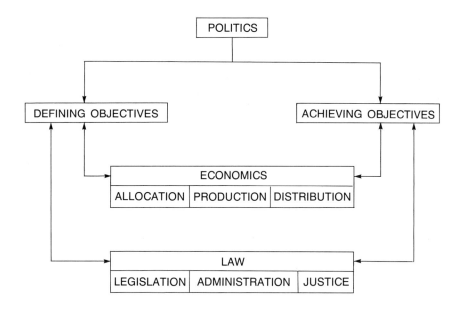

Figure 1. Gene therapy in relation to politics, economics and the law

Gene therapy policy as public policy

The **policy** to be adopted with respect to gene therapy by any government represents public state policy in the general context of health care. When talking about policy in any public sector one usually distinguishes between **setting objectives** and **implementing objectives.** In the area of setting objectives, the competent parliamentary bodies are responsible for integrating the prevailing values in society. A value consensus with respect to gene therapy will have to take into account the existing concepts of public benefit of such therapy, the basic ideologies of a given country and the problems which can be solved or alleviated through gene therapy. Any **value consensus** in a parliamentary discussion will, however, not just be governed by factual considerations, but will also be influenced by party politics, by public and published opinion, and by re-election opportunities as perceived by various political constituencies.

Once the objectives with respect to availability, use and control of gene therapy have been set, the government will also be responsible for their **implementation.** Public health-care policy implies the exercise of authority in carrying out parliamentary mandates. The measures to be implemented will depend to a large extent on their expected efficiency and probability of success, as well as on available financial and other resources. Also, any policy implementation is influenced by bureaucratic elements such as attitude to risk, budgetary considerations and the like.

165

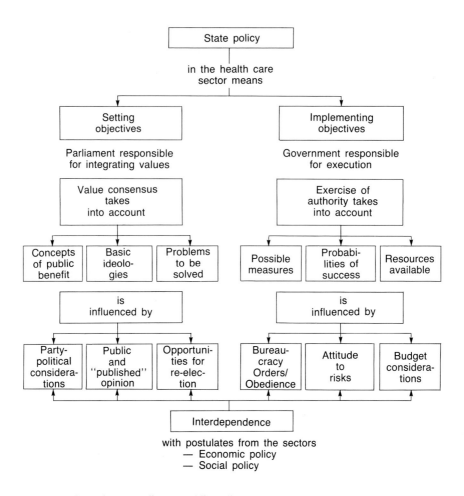

Figure 2. Gene therapy policy as public policy

The setting and implementation of gene-therapy policy will thus have to take into consideration the prevailing societal values as well as the practical realities of public health-care in general (Fig. 2).

Gene therapy and the magic pentagon of health policy

Health policy in general must be regarded as the result of various basic policy considerations which relate to equality of opportunity in health care, quality of performance, compatibility with existing needs, economic efficiency and financial feasibility (Fig. 3).

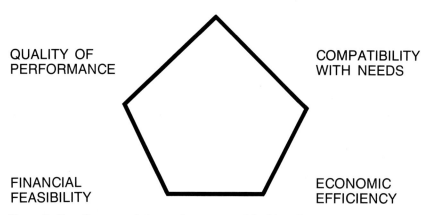

Figure 3. Gene therapy and the magic pentagon of health policy

Equality of opportunity

Equality of opportunity with regard to access to all goods and services provided under the health care system, i.e. ensuring equitable distribution, utilization, and availability of gene therapy, would have to be a fundamental objective of health policy.

Quality of performance

How well a health care system is functioning can be judged by how quickly and efficiently causes of disease and risk factors are recognized and combated, to what extent illnesses are diagnosed and treated at an early stage, how successfully physically or mentally handicapped persons are able to take up work and be integrated into society, and how satisfactorily persons in need of care are looked after. Gene therapy could play an increasingly important role with respect to the quality-of-performance element in health care.

Compatibility with needs

It is now generally recognized that the form and scale of the medical services provided should be more closely adapted to the needs of the individuals concerned. These needs — which are not necessarily identical with the demands being made on the health care system or commensurate with its ability to meet them — are particularly influenced by changes in morbidity and in definitions of what constitutes illness, by changes in risk factors, in patterns of behaviour affecting health, and in standards of popular education, and by changes in the structure and development of the population, as well as by medical and technological progress. It will be of prime importance to define clearly the ''needs'' for gene therapy in both objective and subjective terms.

Economic efficiency

To ensure that it remains possible to finance the health care system in the long term, and thus to maintain and improve the quality of its performance, it is absolutely essential that the means and resources available be exploited economically. It is not sufficient to apply considerations of economy to certain specific institutions and services of the health care system. The relationship between duties defined and services actually rendered, as well as between funds expended and positive effects achieved in practice, must be examined and assessed not only in the various branches and institutions of the health care system, but also, as far as possible, in the entire system. Gene therapy could play a major role in making health care systems more efficient.

Financial feasibility

During past years the costs of health care systems have risen more steeply than national income or gross national product. Marked increases in expenditure incurred by health insurance organizations have resulted in a heavier drain on the incomes of the persons insured, and in business enterprises also having to meet increasing costs. Who should have access to gene therapy and how the resulting costs will be financed will be a subject of intense deliberation.

Factors influencing human health

Human health can be influenced by the environment, by individual behaviour and by pre-existing biological factors.

Influences from the outside emanate from the environment at large and from the society in which a person is living. The risks to human health to be dealt with are primarily those over which individuals have little or no control, because they do not perceive them or cannot avoid them.

Influences from within the individual are those that result from personal habits, such as diet, consumption patterns, amount of exercise taken, or from lifestyle attitudes, type of occupation, leisure pursuits etc. The risks to be dealt with are primarily those over which individuals have personal control, because they either can perceive them or can avoid them by personal action and responsible behaviour.

In addition to the environmental and behavioural influences on health there are also biological pre-existing factors, which affect individual health status. Inherited constitution, genetic structure and predisposition, as well as inherited congenital abnormalities with physical on mental manifestations, are the risk factors with which society has to cope in an optimal fashion. Gene therapy could become a useful tool in dealing with the biological factors which influence human health (Fig. 4).

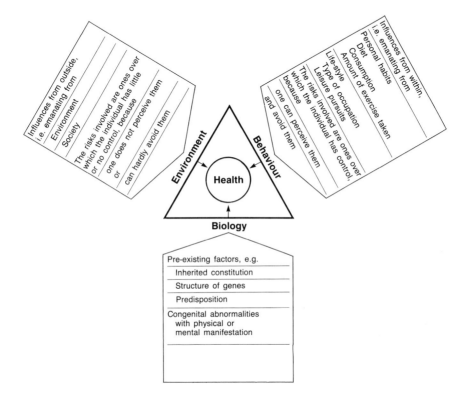

Figure 4. Factors influencing human health

The three main areas in which the State can contribute to the safeguarding of health

If human health is primarily influenced by the environment, by personal behaviour and by pre-existing biological factors, then the government can contribute significantly to the safeguarding of health by measures which relate to a better environment, to improved individual health-related behaviour, or to a betterment of biological factors.

As far as the government's role with respect to the environment and society is concerned, numerous measures are possible to shape the environment and society in the interests of health protection.

With respect to individual behaviour, the government has a variety of intervention possibilities at its disposal to influence health-related behaviour in the interests of better health promotion.

As far as biological factors are concerned, the primary role of government up to now has been to provide an adequate supply of medical goods and services commensurate with the needs of the public in the interests of assuring **health care**. The measures available for that purpose usually are the financing and control of supply of, access to and demand for medical goods and services, and the licensing of health-care professions and specific health industries.

With the advent of the possibility of gene therapy new avenues are opening up towards improving public health care by early measures of intervention. To what extent such measures are necessary and appropriate will ultimately have to be decided in the previously discussed context of setting public health policy (Fig. 5).

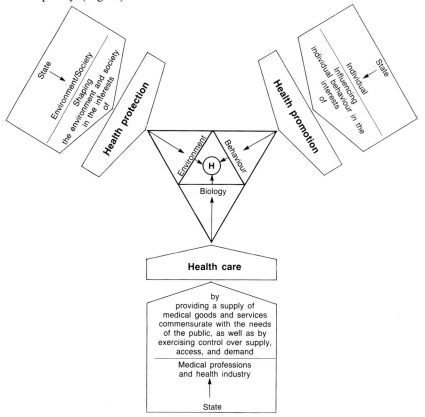

Figure 5. The three main areas in which the State can contribute to the safeguarding of health

170

Government and the function of health protection

In the area of **health protection** the primary objective of government measures is to protect people from health risks that would lead to diseases that cannot be prevented by responsible personal action.

Government measures of this nature afford protection against health impairment attributable to impure food, and water and air pollution. Rules and regulations are also necessary to protect society from harmful emissions at places of work, as well as in the particular and in the general environment.

Typical health-protection measures are also taken by government to combat certain types of communicable or widely prevalent diseases. They include especially:

— **measures to combat communicable or widely prevalent diseases**
   Examples:
      Compulsory immunization/quarantine
      Direct subsidies for the control of certain diseases

— **measures to afford protection against health impairment attributable to food, water, air pollution, drugs etc.**
   Examples:
      Controls imposed on foodstuffs
      Water conservation
      Controls imposed on pharmaceuticals

— **measures to afford protection from harmful emissions at places of work, as well as in the particular and in the general environment**
   Examples:
      Industrial medicine
      Noise abatement
      Protection against radiation
      Protection of the environment

Governmental measures related to health promotion

In addition to protecting the public from outside environmental health risks, the government has a mandate to legislate in the area of **health promotion**. In that category fall all efforts made by the government to influence patterns of human behaviour with a view to preventing disease. The measures are intended to stimulate in people a response which would lead to the avoidance of health risks or diseases due primarily to a lack of health-conscious individual behaviour.

Typical health-promotion measures are governmental activities to encourage a healthy lifestyle, active government intervention to contain non-communicable diseases, or governmental measures to ensure early diagnosis of disease, so that prompt treatment can be given.

Genetic diagnosis opens up a new field of governmental health promotion, which needs to be carefully evaluated. Early genetic diagnosis of later-occurring diseases appears to make sense, especially if the natural course of the expected disease can be altered by means of early treatment or changes in lifestyle or dietary habits. The aspect of confidentiality poses difficult problems to be dealt with prior to allowing any mandatory genetic testing.

Health promotion measures consist of any type of effort made by the State to influence patterns of human behaviour with a view to preventing disease, including especially:

— **measures to encourage a healthy lifestyle**
    Examples:
        Educational campaigns on health risks
        Education designed to promote good health
        Advisory services for sectors of the population at risk

— **measures to prevent non-communicable diseases**
    Examples:
        Iodisation of cooking salt for the prevention of goitre
        Fluoridation of drinking water for the prevention of dental caries

— **measures to ensure early diagnosis of disease, so that prompt treatment can be given**
    Examples:
        Mass X-ray campaigns
        School medical services
        Encouragement to undergo regular medical check-ups
        Genetic screening

Governmental measures related to health care

Health care measures carried out by governments comprise all types of state control, regulation, and guidance of institutions, persons, products, services and market conditions in the interests of providing adequate medical care.

It is usual to distinguish between **measures affecting outpatient treatment** by the licensing and control of medical and ambulatory services, and **measures affecting inpatient hospital treatment** by means of control and supervision of public and private hospitals, as well as **measures affecting the supply of goods** from the health industry through regulation of supply and conditions.

Once gene therapy is established in the armamentarium of medical interventions, it will have to be adequately controlled by governmental health-care measures. Institutions or persons offering gene-therapy goods and services will have to be properly licensed and their professional conduct will have to be subject to appropriate government supervision. Moreover, the financing, pricing and reimbursement of costs of goods and services related to gene therapy will have to be dealt with by government regulation.

172

Health care measures consist of any type of state control, regulation and guidance of institutions, persons, products, services and market conditions in the interests of medical care, including especially:

— **measures affecting ambulatory treatment**
Examples:
Control, regulation, and guidance of supply
Regulating the provision of, and payment for, medical services
Regulating rights and obligations in connection with the doctor-patient relationship

— **measures affecting hospital treatment**
Examples:
Control and supervision of public and private hospitals
Regulating the provision of, and payment for, medical services in hospitals
Regulating rights and obligations of staff and patients in hospitals

— **measures affecting the supply of drugs and the health industry**
Examples:
Control, regulation, and guidance of supply
Regulations concerning prices, reimbursement of drug costs, etc.
Measures serving to regulate demand

Health care measures in the form of primary prevention

A special form of health care consists of **primary prevention measures**. Such government measures are aimed at the prevention of both communicable and noncommunicable diseases. Typical examples are vaccination against poliomyelitis, diphtheria, etc. as well as the iodising of kitchen salt to prevent goitre, or the fluoridation of drinking water to prevent dental caries.

A typical element of such government primary prevention measures in the area of health care is that the **constitutional rights of the individual have to be curtailed** in the larger societal interest. In the case of compulsory chest radiography for the prevention of tuberculosis, for example, individual citizens are required to subject themselves to a form of medical analysis which may be regarded as an invasion of their personal rights. Similarly, compulsory HIV-testing would have to be regarded as an impairment of individual rights, if it were imposed by government regulation. The basic rights of the individual will have to be weighed up carefully against the interests of society at large not to be exposed to infectious diseases. Various legitimate demands and competing value judgements need to be integrated in that process.

The same considerations apply to mandatory genetic testing, as it is becoming standard practice in many countries with respect to so-called "genetic fingerprinting" in criminal investigations.

173

As to gene therapy, it will be very difficult to decide whether any given diagnosed gene deficiency should be dealt with by mandatory gene therapy. **Is there a right to disease in the same way as there exists a right to health?** Who decides on the objective need for gene therapy in any given case? Under what circumstances should gene therapy be allowed when the individual receiving such therapy is unable to make a decision alone? Should parents be allowed to have gene therapy performed in the interest of the yet unborn offspring? These are just a few of the many questions that await an answer following scientific and public debate of the underlying issues.

The question at issue here is that of direct State action based on political decisions and scientific concepts, with the aim of preventing, or reducing the incidence of, certain diseases. Although specific measures to combat communicable diseases really fall within the scope of health protection, state measures to prevent such diseases are often also included under the heading of primary prevention.

Examples of measures for the primary prevention of non-communicable diseases:
Prevention of goitre by iodising kitchen salt
Prevention of dental caries by the fluoridation of drinking water
Prevention of malaria by adding quinine to certain foodstuffs
Prevention of scurvy by adding vitamins to selected basic foodstuffs.

Examples of measures taken by the State to prevent, or reduce the incidence of, communicable diseases:
Frontier check-ups and compulsory quarantine
Compulsory vaccination against poliomyelitis, smallpox, whooping cough, diphtheria, etc.
Compulsory chest radiography for the prevention of tuberculosis.

Health care measures in the form of secondary prevention

The distinction between primary and secondary preventive health-care measures is difficult and sometimes more of a conceptual than of a fundamental nature. One speaks about **primary prevention** with respect to health-care measures aimed at preventing the occurrence of diseases, by means of direct government intervention. Government health care measures in the form of **secondary prevention** are those aimed at the early identification and treatment of diseases that are bound to affect individual health severely if appropriate medical measures are not taken in good time.

Genetic testing offers a whole new field of activity which has yet to be properly determined. The experiences made so far with tests on amniotic fluid in respect of unborn children to detect phenylketonuria, or with specific cancer tests for the early detection of certain forms of cancer, need to be taken into consideration. Tests for the presence of mental disorders or other diseases in which hereditary taints constitute a major factor pose many ethical questions which need further debate.

New promises in the area of gene therapy have to be looked at closely in conjunction with the evolving science and the availability of new forms of genetic diagnosis. Sufficient facilities for **genetic counselling** will have to be established before embarking on a yet unknown future of governmental health care measures dealing with secondary prevention.

Health care in the form of secondary prevention relates to State-regulated activities aimed at the early detection of diseases in the interests of improving the prospects of treatment. Its fields of application include:

— **In unborn children**
  Tests on the amniotic fluid to detect phenylketonuria
— **In infants**
  Early detection of congenital hip dislocation
  Phenylalanine determinations (to detect mental deficiency)
— **In children of school age**
  Early detection of physical handicaps (affecting sight, hearing, speech, gait) by school medical services
  Prompt diagnosis and treatment of mental handicap by school mental health services

Besides the early detection of diseases in children and adolescents, secondary prevention also covers specific serial examinations of populations actually at risk, e.g.:

— **Overweight persons**
  Tests for diabetes
— **Heavy smokers**
— **Women belonging to certain age groups**
  Tests for uterine and breast cancer
— **Persons with hereditary taints**
  Tests for the presence of mental disorders, depression, or other diseases in which heredity is a major factor

Bibliography

Akhurst, R.J. Prospects for Gene Therapy Now and in the Future. *J. Inherited Metab. Dis.* 1989; 12(1): 191-201.

Anderson, W.F. Human Gene Therapy: Why Draw a Line? *J. Med. Philos.* 1989; 14(6): 681-93.

Anderson, W.F. Human Gene Therapy: Scientific and Ethical Considerations. *Recomb. DNA Tech. Bull.* 1985; 8(2): 55-63.

Annas, G.J. *et al.* Legal and Ethical Implications of Fetal Diagnosis and Gene Therapy. *Am. J. Med. Genet.* 1990; 35(2): 215-8.

Bank, A. *et al.* Gene Transfer. A Potential Approach to Gene Therapy for Sickle Cell Disease. 1989; *Ann. N.Y. Acad. Sci.* 565:37-43.

Chang, S.M. *et al.* Prospects for Gene Replacement Therapy. *Birth Defects* 1987; 23(3): 297-321.

Cline, J. Gene Therapy: Current Status and Future Directions. *Schweiz. Med. Wochenschr.* 1986; 166(43): 1459-64.

Cournoyer, D. *et al.* Gene Therapy: A New Approach for the Treatment of Genetic Disorders. *Clin. Pharmacol. Ther.* 1990; 47(1): 1-11.

Dausset, J. Prospects and Ethics of Predictive Medicine. *Pathol. Biol.* 1986; 34(6): 812-3.

Drugan, A. *et al.* Gene Therapy. *Fetal-Ther.* 1987; 2(4): 232-41.

Dworkin, R.B. *et al.* Legal Aspects of Human Genetics. *Annual Rev. Public Health* 1985; pp 107-30.

Fletcher, J.C. *et al.* Ethics Issues in and beyond Prospective Clinical Trials of Human Gene. Therapy. *J. Med. Philos.* 1985; 10(3): 293-309.

Fletcher, J.C. Ethics and Trends in Applied Human Genetics. *Birth Defects* 1983; 19(5): 143-58.

Fowler, G. *et al.* Germ-Line Gene Therapy and the Clinical Ethos of Medical Genetics. *Theor. Med.* 1989; 10(2): 151-65.

Friedmann, T. The Human Genome Project — Some Implications of Extensive "Reverse Genetic" Medicine. *Am. J. Hum. Genet.* 1990; 46(3): 407-14.

Friedmann, T Progress Toward Human Gene Therapy. *Science* 1989; 224: 1275-81.

Gallagher, J. Legal Questions Posed by Gene Therapy. *Am. Assoc. Adv. Scie. Abstr. Pap. Natl. Meet* 1987; 0(153): 43.

Harris, R. The Genetics of Predictive Medicine. *Proc. Annu. Symp. Eugen Soc.* 1983; 19: 149-62.

Hubbard, R. *et al.* Genetic Screening of Prospective Parents and of Workers: Some Scientific and Social Issues. *Int. J. Health Serv.* 1985; 15(2): 231-51.

Jefferson, L. Social and Public Health Implications of Gene Therapy. *Am. Assoc. Adv. Sci. Abstr. Pap. Natl. Meet.* 1987; 0(153): 42.

Johnson, J.M. *et al.* Prenatal Treatment: Medical and Gene Therapy in the Fetus. *Clin. Obstet. Gynecol.* 1988; 31(2): 390-407.

Kinnon, C. *et al.* Somatic Gene Therapy for Genetic Disease. *Arch. Dis. Child.* 1990; 65(1): 2-3.

Knoppers, B.M. Genetic Information and the Law: Constraints, Liability and Rights. *Can. Med. Assoc. J.* 1986; 135(11): 1257-9.

Lange, K. Approximate Confidence Intervals for Risk Prediction in Genetic Counseling. *Am. J. Hum. Genet.* 1986; 38(5): 681-7.

Lappe, M. Prospects for Gene Therapy, Ethical Dimensions. *Am. Assoc. Adv. Sci. Abstr. Pap. Natl. Meet* 1987; 0(153): 43.

Ledley, F.D. Somatic Gene Therapy for Human Disease: Background and Prospects. *J. Pediatr.* 1987; 110(1): 1-8.

Ledley, F.D. Somatic Gene Therapy for Human Disease: Background and Prospects. Part II. *J. Pediatr.* 1987; 110(2): 167-74.

Marx, J.L. The Good News — and the Bad — About Gene Therapy Prospects. *Science* 1987; 236: 29-30.

Moreno, J.D. Private Genes and Public Ethics. *Hastings Cent. Rep.* 1983; 13(5): 5-6.

Müller, H. Human Gene Therapy: Possibilities and Limitations. *Experientia* 1987; 43(4): 375-8.

Nelkin, D. *et al. Dangerous Diagnostics — The Social Power of Biological Information.* Basic Books, Inc., Publishers, New York, 1989.

Newmark, P. Edging Towards Human Gene Therapy. *Nature* 1989; 342:221.

Nichols, E.K. *Human Gene Therapy: The Facts, The Hopes, The Ethical Concerns Surrounding a Revolutionary Treatment of Inherited Disease.* Harvard University Press, Cambridge, Massachusetts, USA 1986.

Orkin, S.H. *et al.* Gene Therapy of Somatic Cells: Status and Prospects. *Prog. Med. Genet.* 1988; 7: 130-42.

Strobel, E. Gene Therapy in the Human. Recommendations of the European Medical Research Council. *Fortschr-Med.* 1989; 107 (4):56.

Walters, L. The Ethics of Human Gene Therapy. *Nature* 1986; 320: 225-7.

Weatherall, D.J. Gene Therapy. *Brit. Med. J.* 1989; 298: 691-3.

Wertz, D.C. *et al.* Ethics and Medical Genetics in the United States: A National Survey. *Am. J. Med. Genet.* 1988; 29(4): 815-27.

Williams, D.A. Gene Transfer and the Prospects for Somatic Gene Therapy. *Hematol. Oncol. Clin. North. Am.* 1988; 2(2): 277-87.

Zabel, B. Gene Therapy. *Monatsschr. Kinderheilkd.* 1989; 147(5): 157-261.

Britain Examines Ethics of Gene Therapy. *New Scientist* 1989; 124: 20 (1989).

Ciba Foundation Human Genetic Information: Science, Law and Ethics. John Wiley & Sons, Chichester, U.K., 1990.

Gene Therapy. *Lancet* 1989; 1: 193-4.

Gene Therapy in Man. Recommendations of European Medical Research Council. *Lancet* 1988; 1: 1271-2.

President's Commission for the Study of Ethical Problems in Medicine and Biomedical and Behavioral Research. Summary from "Splicing Life": A Report on the Social and Ethical Issues of Genetic Engineering With Human Beings. *Recomb DNA Tech. Bull.* 1983; 6(1): 10-12.

REPORTS OF THE WORKING GROUPS

Group A: Human Genome Mapping

Moderator: R.G. Worton[*]
Rapporteur: B.M. Knoppers[**]

State of the art

The Human Genome Project — the international effort to map and sequence the entire human genome, together with the genomes of a few model organisms — is barely four years old. As a formal programme it has only just begun in a few countries, and is still in the planning or discussion stages in many others. If it is carried out according to current predictions, 15 years will be required for its completion: the first five years directed to detailed mapping and related technological development, the second five to the development of new sequencing strategies and data-handling procedures with small-scale pilot projects, and the third five to large-scale sequencing by means of the new technologies.

The project can be divided into three phases: (1) construction of a physical and genetic map with markers every few megabases (Mb); (2) construction of a detailed physical map based on overlapping contiguous cloned segments (contigs); and (3) sequencing of the cloned segments. It is realistic to expect a nearly complete genetic map with resolution of 2-5 centi-Morgans (1 cM is approximately equal to 1 Mb) in the next few years, as some chromosomes are already covered by markers at 1-5 cM intervals. The identification of overlapping YAC (yeast artificial chromosomes) or cosmid clones and the construction of large ($>$ 2 Mb) contigs are proceeding rapidly for a few chromosomes, so the second phase is also expected to be completed for at least some chromosomes in the next five or six years.

The third phase — sequencing — is proceeding more slowly. Less than 0.1% of the human genome has been sequenced. Current procedures are labour-intensive and expensive ($>$ $1 per base pair), making it impractical to use this technology for large-scale sequencing. Computer software to handle the DNA-sequence data is inadequate at present, and new software development is just beginning. Research strategies currently available are suited to limited projects such as sequencing selected genes or gene families (e.g. the HLA region) or of targeted regions of the genome (e.g. the fragile-X site).

---

[*]   Genetics Department and Research Institute, Hospital for Sick Children, and Department of Molecular and Medical Genetics, University of Toronto, Toronto, Canada.

[**] Associate Professor, Faculté de droit, Centre de Recherche en droit public, Université de Montréal, Montréal, Canada.

In traditional medical genetics, screening tests and the identification of disease genes were typically based on knowledge of the protein encoded by the gene. The genetic map generated by the genome project allows new disease genes to be identified, cloned and sequenced without prior knowledge of the protein product. In the next 10 years, almost all the genes responsible for the most common single-gene diseases will be cloned, as will at least a few of the genes that predispose to heart disease, cancer, mental illness and other complex disorders. The identification of genes will in many cases lead to tests for carrier status, to presymptomatic and prenatal diagnosis, and possibly even to new therapeutic strategies.

An international non-governmental body of scientists, known as HUGO (Human Genome Organization), has been formed to facilitate mapping efforts, the exchange of scholars, the organization of international meetings, and the improvement of genetic informatics. Two regional offices — the Americas office in Washington and the European office in London — have been organized, and a third is planned for the Far East. The establishment of local offices is also envisaged. At the governmental level, UNESCO is also playing an important role.

Discussion and conclusions

1. Human genome mapping has much to offer in medical benefits and scientific knowledge and therefore is to be encouraged. Moreover, there is nothing inherently unethical in mapping the human genome. Research on the human genome should be carried out in a concerted and collaborative fashion in the interests of efficient mapping, minimizing overlap, facilitating public awareness of this area of science, and public participation in the formulation of ethical policies.
2. The wisdom of channeling funds into a single targeted research project of the magnitude of the genome project has been questioned. Continued debate is important to determine the relative priority of competing demands among scientific and other activities and to ensure that genome research is funded at an appropriate level in relation to other priorities. To maximize the benefit from the genome effort, research priorities should be periodically re-evaluated.
3. The social and ethical issues raised by the medical application of the knowledge obtained from mapping the human genome are not new, but will be accentuated as the number of identified disease genes increases and as genetic-testing techniques improve and become more widely used. An adequately prepared scientific, medical and lay community is the best defence against misuse of the information and the best assurance that the information and technology will be used in a way that preserves the dignity of the individual.
4. The international nature of the genome effort makes it essential to have international organizations, such as HUGO, UNESCO and CIOMS take an active part in stimulating collaboration and cooperation among nations.

These organizations should accord high priority to the bioethical and social issues associated with human genome mapping and its potential applications.

5. Public and professional education, debate and participation are essential at all levels (children, adults, health-care professionals, community leaders, policy-makers and the media). Various means (mass media, pamphlets, literature and school curricula) should be used to assist in the understanding of the human-genome mapping and of its actual and potential applications. International organizations can do much to facilitate educational programmes and multicultural dialogue about the human-genome project and its implications. Expert panels should be established to be available to respond to requests for explanation from the media and the public.

6. The willingness of individuals and families to make human cells (and the genetic information obtained from them) available for study and banking is a valuable contribution to the human-genome effort. Scientists and physicians have an obligation to disclose fully the nature of the research to be conducted with the cells or tissues, and to obtain voluntary, informed consent. The disclosure should specify the anticipated uses or transfer of both biological material and genetic information. In addition, the privacy of the individual and the family must be preserved by encoding the biological samples and information before their storage and distribution.

7. It is important that all countries benefit from the genome mapping project and participate in the social and ethical discussions, although not all have the resources to conduct genome research or develop new technology. This will require a concerted effort to transfer technology and knowledge from the countries with this expertise to those that are less well-equipped. For the latter, molecular genetic techniques connected with, but not directly dependent upon, genome mapping may be of particular value in the development of simple, rapid and inexpensive identification of pathogenic organisms.

8. The free flow of information derived from the human-genome project is crucial. Information obtained through the use of public monies should come into the international public domain by the normal routes of publication and entry into data-banks. At the same time there is an urgent need for international organizations to address the specific issues of intellectual property, including the interests of the biotechnology industries and researchers. Any such international discussion should include representatives of all the relevant disciplines and interests.

Group B: Genetic Screening and Testing

Moderator: N. Wexler[*]
Rapporteur: R. Kimura[**]

Introduction

Both genetic screening and diagnostic testing refer to the determination of the presence or absence of particular genes or genetic traits. Genetic screening describes testing when it is carried out on a population basis; diagnostic testing is given to individuals who request it. Genetic screening and testing may use the same laboratory tests, although additional assessment may be required for those persons found by screening to have genetic abnormalities. The difference between screening and testing is one of scope: population-based screening is usually large-scale, while diagnostic testing occurs on a case-by-case basis. Screening for carrier status for the sickle-cell trait or Tay Sachs disease, and newborn screening for phenylketonuria (PKU), are examples of population-based screening. There is currently a debate as to whether population-based screening should be offered to detect carriers for cystic fibrosis, as a result of the very recent isolation of that gene.

DNA-based linkage analyses are complex diagnostic tests which must be done on an individual basis. The recently developed DNA linkage test for individuals known to be at risk for developing and possibly transmitting Huntington's disease is an example of individual genetic diagnosis. Genetic screening is often performed in the community (in schools, community centres, or religious institutions, wherever the high-risk population is likely to congregate), but diagnostic testing is usually carried out in a hospital or physician's office.

In the future, many diagnostic and screening tests will be developed on the basis of the findings of the human-genome mapping project. The identification and description of all human genes is a worthwhile pursuit, especially as it will result in programmes being created to translate the findings of the project into genetic services.

The special needs of each nation must be assessed individually to determine the genetic services that will best meet those needs. Most parts of the world have inadequate resources for their pressing needs of basic health services. Some countries may need to give infectious diseases or nutritional deficiencies precedence over genetic testing, while others may be able to provide genetic services for only the most prevalent severe genetic conditions.

---

[*] Professor, Clinical Neuropsychology, College of Physicians and Surgeons of Columbia University, New York, U.S.A.

[**] Director, Asian Bioethics Program, Kennedy Institute of Ethics, Georgetown University, Washington, D.C., U.S.A.; Professor, School of Human Sciences, Waseda University, Tokorozawa, Japan.

A number of international organizations may be able to play a critical catalytic role in assisting nations in preparing plans and programmes for genetic services, to be implemented once a country's more pressing public health needs have been better met.

Epidemiological research will assist in the preparation of a genetic health-needs profile. In some countries, sickle-cell disease is the most compelling genetic problem. For example, in Nigeria, with one-third of the population of Africa, 25% of the population carry a single autosomal recessive gene for the sickle-cell trait and about 1% suffer from sickle-cell disease. In Pakistan, Egypt and other Middle-East countries an educational programme aimed at curbing consanguineous unions may be the most urgent need. The critical point is that programmes must be tailored to the individual needs of each nation and take into account all the other health needs that must be met.

Nations that can develop new genetic technologies should make every effort to devise means of diagnosis and therapy that are inexpensive, accurate and easy to use. These nations should assist in transferring the technology to other nations and continue this assistance until new technologies are in place.

Education

Better education about genetic disorders is urgently needed. Education is required at every level so that health professionals can communicate genetic information in such ways that the public can understand its meaning clearly. Educational programmes should be developed for the following target groups:

1. *Health professionals and ancillary health professionals.* Genetics, including training in genetic counselling, should be part of the curriculum of physicians and other health-care professionals.

2. *Genetic counsellors and paraprofessionals.* Counselling is an essential part of genetic services. Teams of physicians, nurses and master-level genetics associates can constitute an efficient means of carrying out the often extensive counselling and education required for this service to be effective. Often family members and members of voluntary health organizations are capable of educating and counselling, and have personal experience of disorders, which makes the information they impart particularly relevant.

3. *The general public.* Members of the general public know too little about human genetics and genetic disorders to enable them to be informed consumers of genetic services. Genetic education should begin at primary school and continue throughout all levels of education. If individuals are to make informed choices with respect to acquiring and acting on genetic information, they must become familiar with the language of probability and risk. Educational materials, including materials and training programmes for illiterate populations, must *be culturally and educationally appropriate*. Genetic education of the general public is a prerequisite to the mounting of genetic programmes.

4. *Policy-makers.* Those who make public policy must be well informed about the nature of genetic diseases and the availability and scope of genetic services. They should have accurate information, particularly in regard to sensitive areas in which a misunderstanding of genetic disease can have inappropriate or punitive effects. The mistaken equation of a carrier state with a disease may have serious consequences for healthy carriers, who may be discriminated against as if they had the disease. In autosomal recessive conditions, the health of heterozygotes is usually not affected by their having a single copy of the disease gene; in dominant disorders, what is of concern is not the existence of a defective gene but rather the manifestation of the disease, which may be years away at the time of the genetic diagnosis.

Genetic counselling a requirement, not a luxury

There should be equal and affordable access to genetic services within each nation. Nations may differ in their capacity to offer services but the requisite components of a genetics services programme are genetic counselling and prenatal testing (with a possibility of selective abortion). All persons should have free choice among reproductive alternatives and not be coerced by physicians, employers, insurance companies or the State.

Costs should be no barrier in the provision of genetic services. Within the priorities established by each nation, services should be available to all.

Stigma

Genetic information is given to individuals and couples who have their own unique histories and interpret the information in highly personal ways. The information is set in the social context in which the recipients live. Couples will interpret, or misinterpret, genetic information in their own individual ways. Often, recipients of genetic information feel guilty about having passed on an abnormal gene to an offspring or carrying a gene that could lead to disease.

In many cultures, an individual known to have a disease gene is subjected to social stigma, even though the individual may be only a carrier of a recessive gene, and therefore not affected by the gene. Extreme care must be taken in programmes for genetic screening or diagnosis to avoid creating a genetically stigmatized "group", ashamed of their condition and ostracized at home and in the community. In some families, genetic information is not communicated to relatives lest young adults become unmarriageable. The more severe the stigma, the more reluctant are health-care providers to convey genetic information. The diagnosis of a severe disease in a culture in which genetic disease carries a stigma sets the stage for untoward psychological reactions and the possible disintegration of the family. Concentrated public and professional education is necessary to combat such attitudes.

Extreme caution must be taken in developing counselling and testing programmes, particularly for untreatable, incurable, and invariably fatal degenerative disorders such as Huntington's disease and Alzheimer's disease (when a test for Alzheimer's disease becomes available) to ensure that psychological harm is not done to those receiving this devastating information.

Uses and abuses of genetic services

Like all other powerful medical technologies, genetic techniques and services must be used wisely and ethically. Their *use for sex selection (other than to avoid x-linked diseases) is unwise and unethical and should be strongly discouraged.*

Genetic technologies are health-care technologies and must be viewed as such. *The paramount guiding principle in the proper use of genetic services must be concern about an actual or possible health problem.* Extraneous considerations undermine the integrity of the services.

Medical geneticists and allied health professionals should take account of ethical issues and establish proper standards of practice. National and international professional organizations can provide forums for discussing and developing an ethical code of practice.

Confidentiality

Respect for confidentiality must be a governing principle in the development of genetic service programmes. *Genetic diagnostic information must be held in strict confidence.* The only circumstances in which confidentiality might justifiably be breached would be those in which a family member would need to know the genetic information, and the information would be disclosed according to the constraints outlined by the U.S. President's Commission for the Study of Ethical Problems in Medicine and Biomedical and Behavioral Research.

Mandatory versus voluntary testing

Voluntarism should be the guiding principle in the provision of genetic services. Intensive educational campaigns should help to increase service-seeking behaviour. Diagnostic tests can be of the following types:
1. *Prenatal tests.* These tests in particular must be voluntary. Pregnant women should be fully informed regarding tests which may be available and their consequences. It should be their informed choice as to whether or not to be tested.
2. *Screening of newborns.* Screening of newborns should be pursued vigorously, in accordance with standard medical practice in a country. Tests appropriate for each country should be selected, with particular attention to screening for treatable diseases.

184

3. *Testing of carriers for recessive disorders.* Carrier testing for recessive disorders can be carried out at many different stages in the life-cycle, from newborn to adult. Testing at different points in development will entail varying social, economic and medical repercussions. Priority should be given to women of childbearing age and to men of any age who are still procreating, unless earlier screening can be beneficial to health.

4. *Testing of carriers for dominant disorders.* Testing of carriers for dominant disorders is performed to confirm the diagnosis of a genetic disease in an individual or to detect a genetic disease that will appear in the future. Extreme caution must be taken in developing counselling and testing programmes, particularly for severe disorders such as Huntington's disease, or Alzheimer's disease when a test for Alzheimer's becomes available.

Directive as distinct from non-directive counselling

Genetic counselling should be non-directive with respect to the actions to be taken by an individual or couple being counselled. The public must be educated to become informed consumers of genetic services. Even non-directive counselling and voluntary testing can be tantamount to mandatory directive testing when given to a docile or uneducated populace in the context of an authoritarian medical or government system.

Testing for HIV-positivity illustrates some of the complexities of conveying information with profound psychological import. People who test positive often do not take the precautions one would predict, a fact that underscores the futility of directive counselling. The horror associated with the diagnosis can precipitate quite unanticipated and irrational behaviour, emphasizing the need for counselling.

Genetic registers

The establishment of government-sponsored genetic registers may be beneficial to genetic research or to the provision of genetic information to families, but such registers carry the risk of breaching confidentiality. These registers can contain pedigree, epidemiological and demographic data about individuals and families suffering from specific disorders. The setting up of a national register merits careful consideration.

Genetic testing and the work-place

Employers should not require genetic testing as a condition of employment, either to be hired or to continue in employment. When diagnostic tests are indicated they should be given outside the health-care system of the employer, if possible. If they must be given at the work-place, special precautions must be taken to maintain strict confidentiality. Very rarely it may be justifiable to shift to alternative work individuals who are particularly vulnerable, because of their genetic constitution, to a hazard in the environment. When employees are exposed to hazards in the workplace, the employer's responsibility is to make the work-place safe, not to have recourse to genetic testing to discover vulnerable employees.

The primary focus in places of employment should be on the satisfactory performance of work rather than on genetic testing, which may reveal only propensity to certain disorders.

Genetic research on pedigrees

Developing nations have a crucial role in the human genome project, as the mapping of human disease genes depends on having large families. These nations can provide the requisite pedigree resources for mapping. In some instances, these may be large families with a particular disorder; in others they may be inbred families. For example, the family whose collaboration secured the localization of the gene for Huntington's disease lives in Venezela; special circumstances there have permitted the growth and maintenance of this family. In the same manner, cultural, social, and economic factors in the Middle East have encouraged consanguineous unions, which can be studied through homozygosity mapping. Those conducting research in these underdeveloped nations should be attentive to the needs of the local populations. In underdeveloped nations, families participating in research studies may be suffering from a variety of medical and social problems beyond the disease in question. Special precautions must be taken to protect the confidentiality, autonomy and dignity of all individuals and families who give their cooperation for research studies.

Pedigree data obtained from research with families must always be published in a disguised fashion to protect the confidentiality of those whose clinical and genetic information is being shared with the world.

Human rights

Genetic services must be designed in accordance with the principles of the Universal Declaration of Human Rights.

Group C: Human Gene Therapy

Moderator: E. Inouye*
Rapporteur: R. Mullan Cook-Deegan**

Introduction

Gene therapy is the intentional alteration of DNA in human chromosomes to prevent or treat disease. There are two types of target cells. Treatment of *somatic* cells affects only the individual patient. *Germ cell* gene therapy would alter egg cells, sperm cells, or early embryonic cells, so that changes would affect not only the individual but also all or some children of the treated individual.

The debate about gene therapy dates back over 20 years, and has largely reached consensus regarding somatic cell gene therapy. Germ-cell therapy should be broadly discussed before it is upon us, although it is not an immediate prospect. Other issues in genetics — such as genetic screening, genetic testing, discrimination against persons with particular genotypes, and delivery of genetic services — are more pressing. Indeed, gene therapy depends upon a sufficient base of other genetic services for diagnosis and counselling, and can be successful only when these other elements are in place.

At present, gene therapy is a topic relevant mainly to research policy in developed countries. In most parts of the developing world, other health priorities — such as nutrition, child and maternal care, and control of infectious diseases — are far more pressing. Moreover, several of the target diseases in immediate prospect affect only a few individuals even in the developed world, and would be considered insignificant problems in comparison with the diseases affecting millions of people. If gene therapy proves useful in early applications, it must become considerably simpler and less costly before it can be used widely. Special efforts will be needed to encourage collaboration between investigators in developed and developing nations if gene therapy trials are to be expedited, and simple and economical treatments are to be developed, for conditions, such as hemoglobinopathies, that are highly prevalent in the developing world.

The human genome project is related to gene therapy in two respects. The genome project should help locate and elucidate the role of genes that influence disease. Gene maps, DNA clones and probes, analytical instruments

---

*   Science Council of Japan, Tokyo, Japon.

** National Center for Human Genome Research, National Institutes of Health, U.S.A.

and new research methods will clarify gene function and result in new genetic diagnostic techniques. These efforts will not of themselves, however, lead directly to therapy. Whether gene therapy is ultimately possible for a particular disease will depend on physiological and anatomical factors, which will differ markedly among different diseases. As a general rule, the ability to predict disease and understand its pathogenesis will precede the capacity for treatment. The human genome project is important in accelerating the early steps towards gene therapy, and should dramatically enhance research on thousands of genetic diseases, but it does not ensure the success of gene therapy.

To achieve a full understanding of the early natural history of genetic diseases (and certain nongenetic diseases), scientists doing research on some kinds of gene therapy will seek access to cells from embryonic to early fetal stages. The purpose of this research will be to reduce the risks for subsequent subjects of gene therapy. Research involving the embryo and early fetus is highly controversial in many cultures, and this topic merits wide intercultural discussion.

Somatic-cell gene therapy

Somatic-cell gene therapy is technically feasible, and the first approved clinical trials are imminent. Somatic-cell gene therapy will be used initially to treat severely disabling conditions caused by single-gene defects for which current therapies are not fully effective, or to treat other conditions — both genetic and nongenetic — such as cancer and AIDS. If it is successful in such cases, then investigators may contemplate broader application.

Many of the diseases for which gene therapy may eventually be useful are inherited errors of metabolism affecting principally young children. Tay-Sachs disease, for example, has no adult equivalent, and children who develop it are unable themselves to consent to nontherapeutic research. There are provisions for such situations that permit parents to consent when children cannot. If gene therapy becomes a technical prospect, and if preliminary nontherapeutic research is necessary to make it possible, then the possibility of providing a mechanism for such research even in children may be important.

Conclusions

1. Somatic-cell therapy raises no fundamentally new moral issues. For purposes of public policy and clinical application, it should be classed with other innovative therapies. Its risks and benefits should be carefully evaluated by groups independent of the investigators. These groups must review both the technical merit and the social and ethical issues such as patient selection, fair access, informed consent, as well as other issues related to the protection of human beings participating in research and innovative therapy.

2. Gene therapy to treat conditions that cause disability or premature death is acceptable, but doubts arise if it is used for less clearly *medical*

interventions. At some point, alterations cease being medical and become cosmetic. Cosmetic application and alterations intended to enhance normal capacities rather than to correct medical disabilities will require greater discussion. Behavioural and cognitive capacities raise particularly difficult issues. There are some clear cases of disease — such as Alzheimer's disease, Parkinson's disease or Huntington's disease — for which therapy would be welcomed warmly. Attempts to enhance cognitive abilities or behavioural characteristics, however, are inherently suspect, and there is no consensus that such applications are acceptable.

3. Clinical trials of gene therapy in children should be considered only if the need is compelling (e.g., in otherwise untreatable conditions affecting only children) and if the knowledge sought can be gained in no other way.

Germ-cell gene therapy

Many medical treatments and environmental changes have dramatic impacts on future generations, but permanent genetic change is not their primary intended effect. Germ-cell gene therapy would be an exception, because effects on future generations could, in some cases, be the intended result. In other cases, the intention will be to treat an individual for a disease, but with foreknowledge that changes will be inherited.

Germ-cell therapy would be technically more difficult than somatic-cell therapy. First, the inserted gene must be shown not to cause adverse developmental effects. Second, the gene must be incorporated so that it does not cause chromosomal aberrations in subsequent generations. Third, it will probably be more necessary than in somatic-cell therapy for germ-cell changes to target specific chromosomal sites. Progress towards site-directed, controlled genetic modifications may overcome many of the technical barriers to germ-cell therapy.

Although germ-cell therapy is not planned at present, it could conceivably be applied in several clinical situations where somatic-cell treatment would not be effective. Germ-cell therapy might be effective in some kinds of cells, such as muscle cells and neurones, that have ceased dividing and are thus more difficult to approach though somatic-cell therapy, or when several organs must be treated simultaneously. Germ-cell therapy might also be applied soon after fertilization to prevent irreversible damage during fetal development. Except in rare cases, such as offspring of two homozygotes with a recessive disorder, selection of unaffected embryos provides an alternative to germ-cell therapy. The number of clinical situations for which germ-cell therapy would be preferred thus seems narrow; the technical prospects for germ-cell therapy can be evaluated only at some future date when it is contemplated in a concrete clinical situation.

Conclusions

1. Although germ-cell gene therapy is not contemplated at present, continued discussion of germ-cell gene therapy is nonetheless important. The

option of germ-cell gene therapy must not be prematurely foreclosed. It may some day offer clinical benefits attainable in no other way. Science has confounded many predictions about what is technically possible and what is not. Germ-cell therapy might eventually permit more effective prevention of genetic disease, rather than treatment of its effects.

2. Before germ-cell therapy is seriously considered, it must demonstrate its long-term safety beyond other standard treatments, since changes would affect not only the patient but also those progeny who inherit the chromosome bearing the inserted genes. Benefits, however, would also be propagated into future generations. There would have to be confidence that, when treatment affecting future generations is undertaken, descendants of those so treated would still agree with the decision generations later.

3. Germ-cell gene therapy introduces several factors not generally considered when evaluating somatic-cell gene therapy and other innovative therapies. Its benefits extend beyond the treated individual. This raises the possibility of helping future progeny in addition to the patient at hand. The risks of such intervention, however, also extend into future generations, and complicate the evaluation of risks. The function of genes that confer unknown but nonetheless important benefits could be inadvertently modified by germ-cell manipulations, or harmful genes might be propagated irreversibly into the population. These factors must be explicitly evaluated before approval of the first trials of germ-cell gene therapy.

4. The prospect of benefiting future generations could lead to policies that subordinate the interests of the patient at hand to the interests of future generations, or of "the gene pool". The interests of individuals should not be subject to coercive measures aimed at improving the health of the population in general.

General conclusions

1. Gene therapy is a topic of concern worldwide. While the acceptability of somatic-cell treatment of severe diseases is now internationally recognized in a variety of cultures and among widely diverse religious groups, there remain unresolved questions. There is no consensus about whether germ-cell gene therapy in human beings should ever be attempted, and there is a strong need for further debate. There is also a need to ensure public participation and transcultural discussion of how broadly somatic-cell gene therapy should be applied. An international committee, also concerned with other issues in genetic medicine, would be one means to this end, serving as an appropriate forum for discussion of both somatic- and germ-cell gene therapy.

2. The broad social implications of genetics and genetic services, including effects on extremely private and central aspects of human life, make clear the need for training in bioethics among those working in these fields. Organizations concerned with education in these and related fields should consider how best to incorporate bioethics into educational programmes.

REFLECTIONS ON THE CONFERENCE
A Scientist's and a Physician's Perspective

H. Danielsson* and B.O. Osuntokun**

The conference has provided a number of valuable, incisive and knowledgeable contributions by participants who represent an impressive cross-section of science, medicine, philosophy, religion and policy-making.

The opening addresses emphasized the far-reaching consequences of the human genome project, in increased knowledge and in improved methods of diagnosis, treatment and prevention of disease. They stressed also the need for public support of the project and of public participation in dealing with the application of the knowledge which it will generate.

The need for the human genome project

The conference agreed on the great need of the project. We share this view. The mapping and sequencing of the human genome will provide new tools for the prevention, early diagnosis and treatment of such widespread health problems as cardiovascular disease, cancer and diabetes, as well as the 4,400 or more single-gene-defect diseases and the many more genetic defects that bring about inherited susceptibility to disease. Application of the knowledge and ensuing technology will contribute to the control and treatment of tropical diseases that afflict more than a quarter of the world's population. Knowledge gained from the project will improve our understanding of how man functions; it will establish the commonality of the human race and provide a new basis for medicine in the coming decades. The outcome of the project has commercial implications of interest and value to health — new drugs and vaccines, new methods of treatment, new and more simple and reliable methods of diagnosis, and new possibilities for prevention.

The fears of those who oppose the project are well known — notably the fear that funds will be diverted from other important research projects. Others include the unethical use of the new knowledge by insurance companies, employers and governments. The benefits expected from the project in science, medicine and public health, and in developed and developing countries, should be clearly spelt out.

The human genome project raises no new ethical considerations. Its results lead to more accurate instruments than any previously available. Existing ethical guidelines do not have to be fundamentally changed, but rather made more precise to ensure justice, equity, compliance, confidentiality and privacy, and to prevent abuse.

* Secretary, Swedish Medical Research Council, Stockholm, Sweden.

** Professor, Department of Medicine, University of Ibadan, Ibadan, Nigeria.

191

There is a great need to educate the public and the professions, and to provide information to stimulate meaningful public debate, about the project.

Human genome mapping

The formal presentations dealt with the scientific, ethical and human-values, and policy-making aspects of the project.

The major issues are those of project control and coordination, cost and international collaboration, and sharing of data as well as of costs. Ethical issues, apart from those that are well known, include the need for global responsibility for the project and involvement of the developing countries so that they can benefit from the knowledge and technology it will generate. As stated by Watson and Cook-Deegan: "The human genome is inherently international. It necessitates a coordinated world-wide effort to spread the burden of funding the research and to take advantage of the unique genetic resources that can be found anywhere in the world. The nations of the world must see to it that the human genome belongs to the world's people, as opposed to its nations." In our view it would be unethical not to ensure that global involvement is genuinely pursued, and that developing countries are assisted to develop resources (personnel and infrastructure) for participation in the project to contribute to, as well as to apply, the knowledge. It is worth noting that UNESCO proposes to support fellowships and to consider establishing regional centres for genetic research in Africa, Latin America and Asia. It is easier to map rare recessive characters in communities with high rates of consanguineous marriages, as in some populations in the Middle East, South-East Asia and Africa, than in Western populations. The human genome project will provide tools to study such phenomena, but they will require well-trained local scientists.

International cooperation is necessary so as not to duplicate efforts unnecessarily, and to facilitate the free and open exchange of data, ensure that ethical rules are kept, and update data bases and make them accessible world-wide. With respect to duplication, it must be emphasized that, as in other biomedical research, confirmation of one laboratory's results by other laboratories is needed in many, probably most, instances. As to organizational aspects, the question can be asked whether HUGO can fulfil the role or whether other organizations such as WHO and UNESCO need to be involved. CIOMS seems admirably placed to be an ethical watchdog. HUGO is supported by nongovernmental organizations. Should governments of countries committed to the project not be involved? We believe that HUGO is sufficient at the organizational level, but we propose that it should include representatives of the developing countries.

On the issue of whether the project should be implemented by specialized centres under national control rather than by means of investigator-initiated projects, although it might appear that more progress could be achieved by specialized centres, would their exclusive use not have the effect of unduly restricting the availability of funds for investigator-initiated research? At this stage, we believe that much interesting work can be done at the investigator-initiated level and that it is therefore too early to contemplate national or international centres.

Genetic screening

In this section, papers were presented on thalassaemia, Huntington's disease and amyloidosis, and on the ethical, human-values and policy-making aspects of genetic screening.

In translating the products of the human genome project into services, both developed and developing countries, because of limited resources, must take account of cost-effectiveness and cost-benefit. All countries, whether developed or developing, should benefit from the new knowledge, which would lead to extended services in genetic screening. In developing countries, this cannot happen unless there is increased capability to determine the epidemiological profile of genetic diseases, by the use of appropriate resources, and thus determine priorities. Developing countries may need external assistance to supplement national commitment. It should be emphasized that measures to reduce the burden of genetic diseases may not require sophisticated technology — for example, education of the public on the risks of consanguineous marriages may be as effective as any form of genetic screening in some countries in the Middle East, South-East Asia and Africa.

Developing countries should participate in the generation and application of knowledge derived from the human genome project. For this purpose, training is necessary.

For ethical reasons, equity and accessibility of services should be guaranteed, as far as possible, according to national resources and priorities.

We would propose that WHO develop plans for genetic services, particularly for counselling, in developing countries; the plans of a given developing country should be tailored to its needs and resources. WHO would also play an important role in facilitating technology transfer to developing countries.

Finally, we would stress that genetic tests should not be mandatory. However, there is a great need to educate the public so that it comes to realize the value of genetic tests.

Gene therapy

In this section scientific aspects of the treatment of adenosine deaminase deficiency and Lesch-Nyhan's syndrome were described. As in the other sections the ethical, human-values and policy-making aspects of gene therapy were presented.

First, we would refer to the guidelines for gene therapy issued by the European Medical Research Councils (Lancet 1988, Vol. 1, 1271) and the Canadian Medical Research Council in 1990. We consider these guidelines very appropriate. As we see it, gene therapy is, and will be for the next few years, at the research stage. We strongly support the continuation and strengthening of such research.

The conference discussed at some length the pros and cons of germ-line gene therapy. We consider that such therapy raises much more difficult technical as well as ethical problems than does somatic-cell gene therapy, and that the time is not ripe to consider germ-line gene therapy seriously.

In conclusion, the Round-Table Conference has shown in an admirable way the value of the CIOMS series of conferences on Health Policy, Ethics and Human Values — an International Dialogue.

CLOSING OF THE CONFERENCE

F. Vilardell
President, CIOMS

Professor Yagi,
Members of the Organizing Committee of the XXIVth CIOMS Round Table Conference,
Members of CIOMS and Distinguished Guests,

May I say just a few words, as President of CIOMS, to bring to a close this Conference on Genetics, Ethics and Human Values. The subject is a very difficult one and, as customary in CIOMS meetings, discussions have been illuminating, thanks to the multidisciplinary and transcultural background of the participants, who have been able to bring together a variety of beliefs and opinions in a highly intelligent fashion, which I like to think, is the trade mark of our Organization.

I should like to thank the moderators and rapporteurs of the three working parties. Most especially, I should like to acknowledge the unfailing leadership provided by Professor Capron, who has been able to integrate into a whole the various separate issues and their discussions.

The tremendous help provided by the Chairman of the Conference, Professor Yagi and his staff, in particular Dr Yoshi, cannot be overemphasized.

The preparatory work of Dr Bankowski and the CIOMS Secretariat was outstanding, as was the planning of the Conference by the Programme Committee.

As I have already stated on many occasions, may I confess that I have been attending CIOMS meetings for many years because I believe that, by doing so, I educate and improve myself and contribute to the welfare of my patients. I shall return home with the very positive feeling that I have done so again.

May I add a strong plea for including ethical issues in the curriculum of all teaching institutions training personnel in various aspects of genetics.

Thank you again, Professor Yagi, for your perfect and unfailing hospitality.

Thank you all for your dedicated collaboration.

The XXIVth CIOMS Conference is now officially adjourned.

LIST OF PARTICIPANTS

Abe, T. Teikyo University, 11-1, Kaga 2-chome, Itabashi-ku, Tokyo 173, Japan.

Al-Bustan, M. Faculty of Medicine, Kuwait University, P. O. Box 24923, Safat, 13110 Safat, Kuwait.

Aoki, K. Aichi Cancer Center, 1-1, Kanokoden, Chikusa-ku, Nagoya 464, Japan.

Aoki, K. Life Science Institute, Sophia University, 7-1, Kioi-cho, Chiyoda-ku, Tokyo 102, Japan.

Araki, S. Internal Medicine I, Kumamoto University Medical School, 1-1-1, Honjo, Kumamoto 860, Japan.

Bai, K. Institute of Medical Humanities, Faculty of Medicine, Kitasato University, 1-15-1, Kitasato, Sagamihara-shi, Kanagawa 228, Japan.

Bankowski, Z. Council for International Organizations of Medical Sciences, c/o WHO, Avenue Appia, CH-1211 Geneva 27, Switzerland.

Bhamarapravati, N. Mahidol University, 198/2 Trok Wat Saowakhon, Bang Yi Khan, Bangkok-Noi, Bangkok 10700, Thailand.

Binamé, G. Groupe Parlementaire PSC-Chambre des Représentants, Maison des Parlementaires, Palais de la Nation, Rue de Louvain 21,1000 Bruxelles, Belgium.

Bonne-Tamir, B. Department of Human Genetics, Tel Aviv University, Sackler School of Medicine, Ramat Aviv 69978, Israel.

Boulyjenkov, V. Hereditary Diseases Programme, WHO, Avenue Appia, CH-1211 Geneva 27, Switzerland.

Cao, A. Institute for Clinical Biology and Evolution, University of Cagliari, via Jenner s/n, I-09100, Cagliari, Italy.

Capron, A. M. The Law Center, University of Southern California, University Park, Los Angeles, California 90089-0071, U.S.A.

Chalaby-Amsler, K. Council for International Organizations of Medical Sciences, c/o, WHO, Avenue Appia, CH-1211 Geneva 27, Switzerland.

Chijiya, M. Life Science Division, Science and Technology Agency, 2-2-1, Kasumigaseki, Chiyoda-ku, Tokyo 100, Japan.

Cohen, G. N. Laboratoire d'Expression des Gènes Eucaryotes, Institut Pasteur, rue du Dr. Roux, 75724 Paris, France.

Cook-Deegan, R. M. National Center for Human Genome Research, National Institutes of Health, Bethesda, Maryland, U.S.A.

Danielsson, H. Swedish Medical Research Council, P. O. Box 6713, S-113 85 Stockholm, Sweden.

Day, S.B. International Health University of New York Medical College, 6 Lomond Avenue, Spring Valley, New York 10977, U.S.A.

El-Gindy, A. R. Islamic Organization for Medical Sciences, P.O. Box 31280, Suleibekhat, Code 90803, Kuwait.

Fernando, M. 13 School Lane, Nawala, Rajagiriya, Sri Lanka.

Fletcher, J.C. Center for Biomedical Ethics, Box 348, University of Virginia, Charlottesville, Virginia 22908, U.S.A.

Fraser, C. Royal Commission on New Reproduction Technologies, P.O. Box 1566, Postal Station 'B', Ottawa, Ontario KIP 5R5, Canada.

Friedmann, T. Center for Molecular Genetics, Department of Pediatrics, University of California San Diego School of Medicine, La Jolla, California 92092-0634, U.S.A.

Fujii, T. Nagoya Universtiy, Furo-cho, Chikusa-ku, Nagoya 464, Japan.

Fujiki, N. Department of Internal Medicine and Medical Genetics, Fukui Medical School, Fukui 910-11, Japan.

Gevers, J.K.M. Health Law, Institute of Social Medicine, University of Amsterdam, Academic Medical Centre, Meibergdreef 15, NL-1105 AZ Amsterdam, Netherlands.

Gregorios, P.M. Delhi Orthodox Centre, 2 Institutional Area, Tughlaqabad, New Delhi 110062, India.

Hachen, J. Geneva University Hospital, International College of Angiology, P.O. Box 249, 1211 Geneva 25, Switzerland.

Hamaguchi, H. Department of Human Genetics, Institute of Basic Medical Sciences, University of Tsukuba, Tsukuba, Ibaraki 305, Japan.

Hidaka, Y. Department of Internal Medicine, University of Michigan Medical School, 5520 MSRBI, 1150 West Medical Center Drive, Ann Arbor, Michigan 48109, U.S.A.

Ikawa, Y. Laboratory of Molecular Oncology, The Institute of Physical and Chemical Research, 2-1, Hirosawa, Wako, Saitama 351-01, Japan.

Inayatullah, A. Family Planning Association of Pakistan, 3-A Temple Road, Lahore-54000, Pakistan.

Inokuchi, K. Saga Koseikan Hospital, 12-9 Mizugae 1-chome, Saga 840, Japan.

Inouye, E. Science Council of Japan, 22-34, Roppongi 7-chome, Minato-ku, Tokyo 106, Japan.

Ishikawa, T. School of Science and Engineering, Teikyo University, 1189-4, Nagaokacho, Utsunomiya 320, Japan.

Iwatsubo, E. Iizuka Spinal Injuries Center, 550-4, Igisu, Iizuka 820, Japan.

Kagawa, Y. Biochemistry I, Jichi Medical School, 3311-1, Yakushiji, Minami-Kawachi-machi, Tochigi 329-04, Japan.

Katsuki, M. Department of DNA Biology, Tokai University School of Medicine, Bohseidai, Isehara 259-11, Japan.

Kawai, T. Department of Clinical Pathology, Jichi Medical School, 3311-1, Yakushiji, Minami-Kawachi-machi, Tochigi 329-04, Japan.

Kemp, M.S. Medical Research Council, 20 Park Crescent, London WIN 4Al, England.

Kimura, R. School of Human Sciences, Waseda University, Tokorozawa 359, Japan; Kennedy Institute of Ethics, Georgetown University, Washington, D.C. U.S.A.

Knoppers, B.M. Faculté de Droit, Université de Montréal, C.P. 6128, Succursale A, Montréal, Québec H3C 3J7, Canada.

Kohl, J. Ministère de la Santé, Direction de la Santé, 57 bd de la Pétrusse, L-2320, Luxembourg.

Kokkonen, P. Department of Administration, National Board of Health, PL 220 (Siltasaarenkatu 18A), SF-00531, Helsinki, Finland.

Kondo, J. Science Council of Japan, 22-34, Roppongi 7-chome, Minato-ku, Tokyo 106, Japan.

Koulischer, L. Service de Génétique, Université de'Liège, Tour de Pathologie B32, B-4000 Liège 1, Belgium.

Kubota, H. National Epilepsy Center, Shizuoka Higashi Hospital, 886 Urushiyama, Shizuoka 420, Japan.

Kuwaki, T. 4-15-16, Minami-Ogikubo, Suginami-ku, Tokyo 167, Japan.

Macer, D. 31 Colwyn Street, Christchurch 5, New Zealand.

Manabe, H. National Center for Cardiovascular Disease, 5-7-1 Fujishirodai, Suita, Osaka 565, Japan.

Matsuda, I. Department of Pediatrics, Kumamoto University Medical School, 1-1-1, Honjo, Kumamoto 860, Japan.

Matsuo, H. Life Science Division, Science and Technology Agency, 2-2-1, Kasumigaseki, Chiyoda-ku 100, Japan.

Mauron, A. Fondation Louis Jeantet de Médecine, P.O. Box 277, CH-1211 Geneva 17, Switzerland.

Milani-Comparetti, M. Istituto Internazionale Studi Etico-Giuridici, Nuova Biologia, Via Cappuccini-Villa Eolian, 98057 Milazzo, Italy.

Miller, J. National Council on Bioethics in Human Research, 74 Stanley, Ottawa, Ontario, KIM 1P4, Canada.

Mishima, S. 3-6-16, Maehara-cho, Koganei, Tokyo 184, Japan.

Miwa, S. Okinawa Memorial Institute for Medical Research, Teranomon Hospital, 2-2-2, Toranomon, Minato-ku, Tokyo 105, Japan.

Mukunyandela, M. Tropical Diseases Research Centre, P.O. Box 71769, Ndola, Zambia.

Muramatsu, M. Department of Biochemistry, Faculty of Medicine, University of Tokyo, 7-3-1, Hongo, Bunkyo-ku, Tokyo 113, Japan.

Müller, H. Swiss Academy of Medical Sciences, Department of Human Genetics, University Children's Hospital, CH-4005 Basel, Switzerland.

Münster, O. The Danish Council of Ethics, 2-4 Ravnsborggade, DK-2200 Copenhagen N, Denmark.

Nagano, K. Department of Biology, Jichi Medical School, 3311-1 Yakushi-ji, Minami-Kawachi-machi, Tochigi 329-04, Japan.

Nakajima, H. Director-General, World Health Organization, 1211 Geneva 27, Switzerland.

Nakatani, K. Kyorin University, 476, Miyashita-cho, Hachiohji, Tokyo 192, Japan.

Ofstad, J. The National Committee of Medical Research Ethics, Haukeland Hospital, Medical Department A, 5021 Bergen, Norway.

Ogasawara, N. Institute for Developmental Research, 713-8, Kamiya-cho, Kasugai, Aichi 480-03, Japan.

Ogawa, Y. Department of Pediatrics, Saitama Medical Center, 1981, Kamoda, Tsujidomachi, Kawagoe 350, Japan.

Ohkura, K. The Genetic Counseling Centre of the Japan Family Planning Ass., Inc. c/o Hoken-Kaikan, Ichigaya, Sadohara-cho 1-2, Shinjuku-ku, Tokyo 162, Japan.

Okamoto, M. Prime Minister's Office, Council for Science and Technology, Science and Technology Agency, 2-2-1, Kasumigaseki, Chiyoda-ku, Tokyo 100, Japan.

Osuntokun, B.O. Department of Medicine, University of Ibadan, Ibadan, Nigeria.

Palella, T.D. Department of Internal Medicine, University of Michigan Medical School, 3918 Taubman Center, Ann Arbor, Michigan 48109-0358, U.S.A.

Papiha, S.S. Department of Human Genetics, University of Newcastle upon Tyne, 19 Claremont Place, Newcastle upon Tyne, NE2 4AA, England.

Parkman, R. Division of Research Immunology and Bone Marrow Transplantation, Childrens Hospital Los Angeles, 4650 Sunset Boulevard, P.O. Box 54700, Los Angeles, California 90054-0700, U.S.A.

Roberts, D.F. Department of Human Genetics, University of Newcastle upon Tyne, 19 Claremont Place, Newcastle upon Tyne, NE2 4AA, England.

Saiki, R.K. Department of Human Genetics, Cetus Corporation, 1400 Fifty-Third Street, Emeryville, California 94608, U.S.A.

Sakaki, Y. Research Laboratory for Genetic Information, Kyushu University, 3-3-1, Maedashi, Higashi-ku, Fukuoka 812, Japan.

Sano, K. Department of Neurosurgery, Teikyo University School of Medicine, 2-11-1, Kaga, Itabashi-ku, Tokyo 173, Japan.

Shimizu, N. Department of Molecular Biology, School of Medicine, Keio University, 35, Shinano-machi, Shinjuku-ku, Tokyo 160, Japan.

Shirai, Y. Faculty of Arts, Shinshu University, 3-1-1, Asahi, Matsumoto 390, Japan.

Sugino, Y. Takeda Chemical Industries Ltd., 3-6, Doshomachi, 2-chome, Chuo-ku, Osaka 541, Japan.

Takagi, K. The House of Councillors, Diet of Japan, 2-1-1, Nagata-cho, Chiyoda-ku, Tokyo 100, Japan.

Takao, A. The Heart Institute of Japan, Tokyo Womens' Medical College, 8-1, Kawadacho, Shinjuku-ku, Tokyo 162, Japan.

Tanaka, M. Department of Biomedical Chemistry, Faculty of Medicine, Nagoya University, 65, Tsuruma-cho, Showa-ku, Nagoya 466, Japan.

Tharien, A.K. Christian Fellowship Hospital, Oddanchatram-624619, Anna District, India.

Tsuyama, N. Japanese Society for Rehabilitation of the Disabled, 20-8, Nishishinjuku, 7-chome, Shinjuku-ku, Tokyo 160, Japan.

Uematsu, T. 1-13-21, Nishiogkikita, Suginami-ku, Tokyo 167, Japan.

Uusmann, I. Ministry of Health and Social Affairs, S-103 33 Stockholm, Sweden.

Verma, I.C. Genetics Unit, Department of Pediatrics, Old Operation Theatre Building, All India Institute of Medical Sciences, Ansari Nagar, New Delhi 110 029, India.

Vilardell, F. Council for International Organizations of Medical Sciences, Juan Sebastian Bach 11, 08021 Barcelona, Spain.

Vogel, F. Institut für Humangenetik und Anthropologie, Ruprecht-Karls-Universität Heidelberg, Im Neuenheimer Feld 328, D-6900 Heidelberg, Germany.

Von Wartburg, W.P. Department of Law, University of St. Gallen, 9000 St. Gallen, Switzerland.

Wallace, C.W. Tokyo Regional Office, National Science Foundation, Embassy of the United States of America, 1-10-5, Akasaka, Minato-ku, Tokyo 107, Japan.

Wertz, D.C. School of Public Health, Boston University, 80 East Concord Street, Boston, MA 02118, U.S.A.

Wexler, N.S. Department of Neurology and Psychiatry, Columbia University, 722 West 168th Street, Box 58, New York, N.Y. 10032, U.S.A.

Worton, R.G. Department of Genetics, Toronto Hospital for Sick Children, 555 University Avenue, Toronto, Ontario M5G IX8, Canada.

Wyngaarden, J.B. National Academy of Sciences, 2101 Constitution Avenue, N.W.-HA450, Washington D.C. 20418, U.S.A.

Yagi, K. Institute of Applied Biochemistry, Yagi Memorial Park, Gifu 505-01, Japan.

Yamamoto, S. St. Luke's College of Nursing, 10-1, Akashi-cho, Chuo-ku, Tokyo 104, Japan.

Yamashina, I. Department of Biotechnology, Faculty of Engineering, Kyoto University of Industry, Kamigamo-Motoyama, Kita-ku, Kyoto 603, Japan.

Yanase, M.S. Department of Science and Engineering, Sophia University, 7-1, Kioicho, Chiyoda-ku, Tokyo 102, Japan.

Yoshida, M.C. Chromosome Research Unit, Faculty of Science, Hokkaido University, Kita-ku, Sapporo 060, Japan.

Yoshioka, A. Department of Pediatrics, Nara Medical University, 840, Shijo-choo, Kashihara, Nara 634, Japan.